Einstein believed that electro-magnetic (EM) waves (light) were somehow special; they did not require a propagation medium (ether), and that only relative motion between systems was important. A New Relativity theory (NR), an extension of the Lorentz Transform (LT), predicts the same measured properties as Einstein, but based on a propagation medium. It shows that waves and fields, including EM and gravitational, require a medium to propagate their disturbances and make their wave equation causal (predictable). That all motional effects are with respect to the medium, distinguishing between source and observer motion. Gravity appears to be the electrical attraction of difference electric fields from dissimilar charges in atoms and molecules. NR re-unites classical and modern physics and predicts additional measured properties that Einstein's relativity cannot predict.

Order this book online at www.trafford.com
or email orders@trafford.com

Most Trafford titles are also available at major online book retailers.

Printed in the United States of America.

ISBN: 978-1-4669-8042-6 (sc)
ISBN: 978-1-4669-8043-3 (hc)
ISBN: 978-1-4669-8044-0 (e)

Library of Congress Control Number: 2013921694

Trafford rev. 02/22/2014

www.trafford.com

North America & international
toll-free: 1 888 232 4444 (USA & Canada)
fax: 812 355 4082

Unification of Electromagnetism and Gravity

A New Relativity (NR) theory is described in a simple physical way, in an attempt for the lay person to understand. But at the same time updates aspects of modern physics in a rigorous manner. NR, which is based purely on experimental evidence, simplifies our understanding of the universe and challenges relativity as a more logical and comprehensive theory. It is found that motional electromagnetic (EM) and gravitational theories appear to have two inherent deficiencies that have prevented them from becoming a unified theory. (α) Not recognising that the propagation medium (ether) is the essential thread that runs through these developments. (β) Not realising that EM waves and gravity are two forms of the same field - unsteady electric and steady difference electric fields. The new theory re-establishes a preferred frame of reference. It creates links between electrical sources and observers in motion, and between electric fields and gravity. NR predicts the measured aspects of Einstein's Special Relativity (SR) based on a medium, rejects the ether-less aspects of SR that cannot be measured, and predicts additional measured aspects that SR cannot predict. The medium provides the bridge between the Lorentz transform, accelerating frames and gravity, providing a basis for the unification theory of the universe.

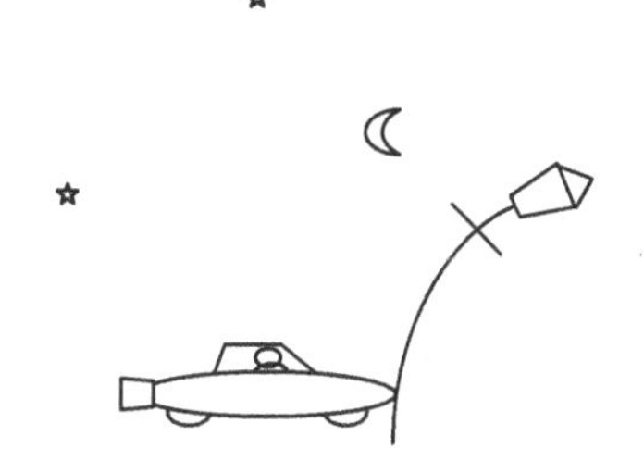

Selwyn E Wright

Solvay Congress 1927. The world's greatest scientists at that time, some of whom play a leading role in this book.

Back Row: A. Piccard, E. Henriot, P. Ehrenfest, Ed. Herzen, Th. De Donder, E. Schrödinger, J.E. Verschaffelt, W. Pauli, W. Heisenberg, R.H. Fowler, L. Brillouin. **Middle Row:** P. Debye, M. Knudsen, W.L. Bragg, H.A. Kramers, P.A.M. Dirac, A.H. Compton, L. de Broglie, M. Born, N. Bohr. **Front Row:** I. Langmuir, M. Planck, M. Curie, H.A. Lorentz, A. Einstein, P. Langevin, Ch. E. Guye, C.T.R. Wilson, O.W. Richardson

TO MY WIFE SHIRLEY

Abstract:

Key words: Lorentz transform, special and general relativity, ether and propagation medium, absolute time and space, time and space travel, unification of electric, electromagnetic, gravitational and inertial fields.

Historically, Lorentz predicted the measured effects of electromagnetic (EM) systems in motion, based on a medium (ether). Einstein partially solved the EM wave equation, based on Lorentz's Transform (LT), it therefore predicts the same medium based properties as Lorentz. However, Einstein claimed that the medium did not exist, predicting additional ether-less non causal properties that have not been measured. A New Relativity theory (NR), based on an extension of Lorentz's LT, distinguishes between source and observer motion with respect to the medium, and between Earth centered and heliocentric medium based reference frames, not considered in Einstein's Special Relativity (SR). Thus, SR has an ether-less, non causal aspect, which cannot be measured, and a medium based aspect that can, but is incomplete. Einstein's measured predictions are medium based, rather than the non causal ether-less predictions he believed.

Having no medium causes insurmountable difficulties: no known way of transmitting EM waves, no way of solving the wave equation, predicting events, maintaining continuity of time and space and no way of creating inertia. Without a preferred frame of reference, wave Propagation Time Asymmetry (PTA), up stream and down, a characteristic of causality, which is clearly observable, cannot be predicted. Time has no direction, one cannot distinguish between which system is moving, the source or observer, which system is ageing least, or identify the optical paths - they are indeterminate. Without a medium, Einstein's ether-less inertial frame results in simultaneity and reciprocity, which have never been measured. Ether-less relativity cannot be supported by the LT or Maxwell's Equations, (ME's), they are both medium based. Einstein's SR based on a medium is causal but incomplete, its ether-less aspects are irrational (non causal).

A causal EM Motional Analysis has been developed, which distinguishes between optical properties of sources and observers in motion and between stationary and moving media. Three pairs of time and space scales (source, observer and medium) are required rather than the two unspecified in SR. The theory distinguishes between measured source (satellite) backward motional angles and observer (telescope) forward aberration angles, not distinguishable in ether-less relativity. It explains the Michelson and Morley experiment, Bradley's stellar aberration and Michelson and Gale's fixed optical loop whose apparent incompatibility led to loss of faith in the medium, in the first place.

A vacuum medium is not empty space, it has a finite and measureable permeability (electrical inertia) and permittivity (electrical stiffness) propagating (bouncing) EM waves through the medium, according to Maxwell. Restoring the medium allows the propagation paths to be determined and to predict casual observations. The complete wave equation allows both source and observer flight paths to be chartered on the same universal medium continuum. This renders material time travel and no absolute time and space as untenable. The medium is generally at rest in space and moves with gravitational bodies (planets). If the planet rotates it forms a medium velocity gradient (boundary layer) above its surface.

Gravity, accelerating frames and the LT are all equivalent to a velocity. Schwarzschild metric describes the medium's attraction and compression by gravity, both time and space. Structures, according to Lorentz, contract passing through the medium, and moving observers relatively expand the medium reducing gravity, thus defining an optical time Equivalence Principle. The electrical medium transmits steady electostatic fields, unsteady electric fields (light), steady difference electric fields (gravity) and a universal residual difference field (creating inertia). Gravity is explained through a weak difference electric field from finite distributions of dissimilar charges (protons and electrons) within atoms and molecules throughout the universe.

Contents

Overview: Unification of Electromagnetism and Gravity

1 Einstein's belief. Einstein believed there was no difference between a stationary and constantly moving (inertial) frame. Or between colliding systems, or motion between a magnet and an attracting metal plate, any could be considered as moving. These apparently indistinguishable motions helped convince Einstein that only inertial frames and relative motion between systems were meaningful. He concluded therefore, that there is no need for a propagation medium (ether). However, relative motion between a source (action or event) and an observer (detector), without a medium, has no meaning in wave theory or the mass-less particle equivalent.

2 Relativists and wave theorists. Relativists (usually astrophysicists) have been content to accept Einstein's view that there is no need for a propagation medium (ether) to propagate EM waves. They use Einstein's Special Relativity (1905) (SR) and General Relativity (GR) (1915) and predict measured results. They see no reason to believe that SR and GR are in error in any way. Whereas, wave theorists find Einstein's ether-less claims as irrational (non causal). There is a fundamental requirement, confirmed by measurement, that all waves need a propagation medium to propagate and to make their wave equation causal (predictable). The EM wave equation can be readily derived and solved using the classical medium based wave equation modified by the Lorentz Transform (LT) (1889), demonstrating the medium's presence and necessity. Not recognising the medium, Einstein's ether-less aspect of SR is unable to satisfy the general wave equation, inabling ether-less SR to distinguish between source and observer motion.

3 Medium rejected. Although Maxwell established the EM medium in 1865, the propagation medium's popularity lost ground in the early 1920's through: i) The Michelson and Morley Experiment (MMX) (1887) revealing no motional effects. ii) Bradley's (1725) stellar aberration appearing to discredit the medium. iii) Failing to distinguish between measured differences between sources and observers in motion. iv) Believing that the medium based Lorentz transform predictions could support Einstein's ether-less relativity. v) Over estimating the relativistic and gravitational effects, which are small compared with the classical motional effect, at Earth speeds, gravitational strengths and short integration times. It appears that all well established experiments attributed to relativistic and ether-less effects: MMX, Bradley, Sagnac (1913), Michelson and Gale (M&G) (1925), Saburi et al (1976), Reasenburg et al (1979) and GPS (1992), are in fact predicted by classical medium based Propagation Time Asymmetry (PTA), not by Einstein's ether-less Propagation Time Symmetry (PTS). M&G, which was regarded as the final demise of medium based theories, is now shown to be medium based.

4 Sense of unease. After a century of non causal ether-less relativity, there will be, initially, a reluctance to accept the propagation medium and abandon Einstein's ether-less concept of relativity. Einstein was convinced that there was no propagation medium. However his thought experiments included ether-less predictions, such as time travel and no absolute time and space, neither of which can be measured. Einstein's reluctance to accept the medium has caused considerable problems that have arisen through the medium's rejection, and numerous accurate predictions based on the medium's presence. Some of us, who have read Einstein's theory of relativity or studied it in some detail, have been left with a sense of unease. Some aspects of the theory appear to be against one's intuition, requiring a leap of faith to suppress a disturbing feeling that the physics is not quite right.

5 Unexplained observations. There should be no obvious unanswered questions; such as how does light propagate, or how does one solve the wave equation for propagating waves, without a propagation medium? Without a medium, there is no physical mechanism to explain how the simple Doppler (1842) effect occurs. How a moving more dense medium than a vacuum can convect light. How an impulsive wave is formed in Cerenkov (1934) radiation. How two systems can move apart physically, greater than the speed of light, but not relative to each other. Or distinguish between light propagation on Earth, around the Earth or through the Solar System and beyond. Finally, Einstein's invariant inertial frame that encouraged his ether-less beliefs, is not in accord with measured medium based predictions.

6 Sky at night. Challenges to Einstein's Special Relativity (SR) (1905) have come and gone. Through dogged defence of the status quo we are still managing to get by with the same theory. Viewing the sky at night it is difficult to imagine that the universe is not continuous and absolute, i.e. it is not one piece of spatial fabric. Einstein concluded just that, implying that space is a patch-work of autonomous regions of relativity. Having no propagation medium, Einstein assumed that there was no absolute time and space. His belief was based mainly on the invariance of the MMX.

7 Michelson and Morley Experiment. The MMX (1887) showed that light on Earth propagated exactly the same as if the Earth was stationary. The propagation time was the same in and against the direction of the Earth moving through space at high speed. For classical waves the time should have been asymmetrical, take longer upstream than down. This was interpreted through Einstein's inertial frame as implying that there was no propagation medium. The establishment, without rigorous proof, also concluded that the universe was without a medium. They, like Einstein, believed they could see the Emperor's new clothes. However, the clothes in retrospect appear to be quite conventional after all. The MMX null result is now shown quite naturally to be based on a medium moving with the Earth.

8 Without a medium. Einstein, in his concept of relativity, believed that only relative motion between space ships was meaningful. Ships could travel notionally at any speed without detection, providing there was no relative velocity between them. This is not a satisfactory situation, how could their actual speeds be measured? According to Einstein's relativity, either ship could be considered to be moving and the other stationary. Either set of astronauts could be considered to age less than those on the other ship. Amazingly, against all logic, both situations were considered possible, even at the same time, which is physically impossible (non causal) in the real world.

9 With a medium. Einstein claimed that the situation could not be resolved until one of the ships changed speed or direction. But what would have been the situation immediately before the ships changed course? What would have happened if the ships had never changed course? These questions only arise through not accepting the medium's presence. Restoring the medium allows these questions to be answered naturally. Accepting the medium's presence ensures finite and definitive velocities relative to the medium, removing all ambiguity. The two ships could, for example, approach each other at speeds just below the speed of light, their total speed relative to the medium being almost twice that of light. But relative to each other, through the relativistic addition of velocities, their speed would always appear to be below that of light.

10 All waves require a medium. To transmit the steady field, its disturbances (waves) and make the predictions causal (solutions of the motional wave equation), a medium is essential. The medium is not a mathematical artefact that can be removed, as Einstein believed. All observed motional effects are caused through the interaction of the moving system with the propagation medium. The medium determines the wave characteristics, including the wave propagation speed. EM waves are no exception; their electrical medium's inertia and compressibility are finite giving a finite speed. If there was no medium its propagation speed would be infinite, which is not the case. Apart from projectiles, there is no other known way of transmitting information and energy across space, unless one believes in some kind of metaphysical method, for which there appears to be no evidence. Historic lethargy has prevented the medium's acceptance for over a hundred years. A similar denial greeted Galileo's insistence that the Earth is not the centre of the universe, four hundred years ago.

11 Inconsistent. Einstein's SR is not consistent with classical wave theory. It implies both ether-less and medium based predictions, which are contradictory. The main problem is that EM waves (light) are supposed to propagate without a medium. Having no medium is irrational, it's against fundamental (causal) physics. This is like water waves without water, or sound waves without air, there is nothing special regarding EM waves. It is established that the medium exists and restoring its presence removes all inconsistencies. It can be shown that Einstein, who tried to convince us of the medium's non existence, actually used a medium in his own field equations, allowing causal predictions to be made, including many of the measured results we are familiar with today. But of course this negates Einstein's own ether-less aspect of relativity.

12 Relativistic Addition of Velocities. The two space ship velocity interpretations are explained quite legitimately through the medium based Relativistic Addition of Velocities (RAV). Each ship retains its individual velocity relative to the medium, which determines its rate of ageing, according to Lorentz. This is the medium at work producing completely rational (predictable) effects. The only non intuitive part of the process is the relativistic addition, which is perfectly rational mathematically, being derived directly from the medium based Lorentz transform. Here the system and light speed is added relativistically, making the light speed 'c' invariant. This is because time and space in the moving frame shrink by exactly the same ratio maintaining its speed. The light speed in the medium, or to a moving observer, then remains unchanged.

13 Vacuum medium. A vacuum medium, without gravitational matter, is not empty space. It has Maxwell's (1865) permeability (electrical inertia) and permittivity (electrical stiffness) propagating (bouncing) electrical disturbances through the medium. It transmits steady electric fields, steady difference electric fields and unsteady electromagnetic fields. These are generated, for example, from electrons jumping orbits in atoms and molecules, radiating light or mass-less photons. Photons, with no rest mass are equivalent to discrete energy bursts of light, another way of representing wave propagation. Photons require a medium just the same as waves and travel at the speed of light relative to the medium.

14 Medium moves with heavy bodies. The medium is generally at rest in space. However, it is attracted to, and through its non-ridged structure moves with gravitational bodies (planets). Here the medium surrounds and orbits with the body within its Gravitational Field of Dominance (GFOD). Stellar aberration, described by Bradley (1725), which was believed would be affected by the medium surrounding and orbiting with the Earth is shown not to be the case. The star light propagating in the medium, 'at rest in space', passes through the orbiting medium, forming the aberration angle in an observing telescope on Earth. This is an actual angle embedded in the medium.

It is not a resolved angle between the Earth's motion and the speed of light that would have resulted if there had been no medium. Filling the telescope with an obvious medium (water), does not affect the stellar angle, confirming the medium's presence. The medium forms a smooth transition (ray bending), between the moving and stationary situations. NR therefore explains the Michelson and Morley Experiment (1887) (medium moves with the Earth), Bradley's (1725) stellar aberration, (telescope moves relative to star light propagating in the stationary medium) and Michelson and Gale's (1925) boundary layer (medium on the Earth's surface moves relative to the surrounding stationary medium), supporting the legitimacy of the entrainment model.

15 Observing the universe. It is self evident that to understand the physics of the universe requires observations, which in turn require light. Observations therefore depend on EM waves and the medium that transmits them. They also depend on motion of the source and observer relative to the medium. Basically, there is nothing extraordinary regarding EM waves, they behave and give similar causal properties as regular classical waves. Familiar wave motions are water waves spreading out from a stone thrown in a pond, waves moving on the surface of the sea or pounding the beaches. They all require a medium to be causal (predictable).

16 Causality. In the case of acoustic waves, for example, the air becomes the medium transmitting sound from its source to the listener. This information is predicted through the medium by solving the classical Navier-Stokes Equations (NSE's). Its solution is causal (the cause must occur before the effect), it is valid for all kinds of wave motion. In the case of EM waves, there is no difference from classical waves, apart from the medium based Maxwell's Equations (ME's) (1865) and the Lorentz Transform (LT) (1899) replacing NSE's. These EM equations predict events (light) propagating through the EM medium, eventually seen by an observer. For stationary EM sources and observers there is no confusion, an EM medium is naturally accepted to solve the wave equation and make the observations causal.

17 Motional problems. However, problems occur when motion is considered. Suddenly, when systems were set in motion, Einstein believed, without proof or robust scientific reasoning, that EM waves were somehow unique, and that a medium was no longer required. This of course is inconsistent and irrational, wave propagation without a propagation medium cannot happen, it is a contradiction. For media movement, relative to a fixed source and observer, or visa versa, the upstream and downstream wave propagation times will become asymmetrical and vary with motion relative to the medium. There is no other rational outcome, it is determined by the causal solution of its wave equation, using a propagation medium

18 Propagation Time Asymmetry. If the water speed of a river, or a shallow stream on the beach, equals the surface wave speed, the waves propagating upstream will stand still relative to their surroundings, giving an infinite upstream and half the downstream propagation time. This simple concept of Propagation Time Asymmetry (PTA) and its variance, applies generally to all kinds of waves. According to fundamental physics there is no wave type that does not satisfy causality and create PTA, including EM waves. This asymmetry, which is indicative of causality, is absent in Einstein's inertial frame.

19 Einstein's Inertial Frame. If an EM system (source and observer of fixed separation) moves at constant speed, relative to the medium, it does not affect the observations. But it does result in PTA according to Lorentz. This variance cannot be accounted for by the ether-less invariant Einstein Inertial Frame (EIF). Einstein's assumption that there is absolutely no difference between a stationary and constantly moving frame is untenable. It is true that the mechanics, propagation speed and ME's are invariant, but the PTA remains variant in the moving frame. Without the medium, EIF cannot propagate waves, it cannot be a solution of the wave equation, it is non-causal.

20 Lorentz fundamental. Lorentz's motional rectangular coordinate axes transform (LT) is the basic theory for systems in motion. It gives measured predictions, based on motion with respect to Maxwell's medium, not relative motion between systems, as Einstein believed. Its predictions are based on the motional solution of the medium based EM wave equation. The effects are therefore predictable (causal). They include the medium based classical PTA, with Doppler (1842) frequency and Sagnac's (1913) time domain asymmetry. In addition to PTA there is the modifying Lorentz's time and space contraction (LC) by high speed system motion relative to the medium. Time and space of the medium do not change, it's the time and structure of physical objects (atoms and molecules) that contract (not dilate) passing through the medium.

21 PTA dominates. Relativists, not accepting the medium's presence, attempt to use relativistic arguments to try to explain basic EM motional effects. However, relativistic effects at Earth speeds and short measurement periods are small compared to the instantaneous classical PTA. Michelson and Morley (1887), Sagnac (1913), Michelson and Gale (1925), Saburi et al (1976) and GPS (1992) are all basically explained using only the medium based classical PTA. The only major difference between classical and EM theories is Lorentz's contraction in the direction of motion at high speed. Both theories exhibit the classical medium based PTA, in the moving frame, it is the vital (dominant) part of any causal motional wave theory.

22 Rectangular axes. Relativists attempt to argue that no propagation medium exists or indeed is required. They interpret Minkowski's (1908) rectangular axes space-time four vector analysis, which plots moving systems in space and time, as requiring no medium. However, the vertical and horizontal axes of the space-time diagram use Lorentz's rectangular axes medium based transform, representing time and space respectively, normalised against the speed of light. The velocity of light then becomes a 45 degree gradient. This is just a mathematical convenience of representing time and space. *There is no mechanism or physical justification to remove the medium.*

23 Oblique axes. In an effort to remove the effect of the medium and support Einstein's concept of relative motion, Lorentz's medium based rectangular axes transform can be replaced by oblique axes, according to Born (1924). These axes were an attempt, using Minkowski's space-time, to simulate simultaneity (propagation time symmetry upstream and down) and reciprocity (interchanging sources and observers made no difference to the observations). However, these ether-less properties, where the medium's absence is argued using medium based concepts (circular arguments), are non causal, they are not a solution of the wave equation. They imply time travel and no absolute time and space, neither of which have been measured.

24 Ether-less relativity. Credibility is raised by the fact that no one has ever verified Einstein's ether-less relativity, nor established the redundancy of the propagation medium. Researchers claiming to have verified ether-less aspects of SR have usually verified the medium based Lorentz Transform (LT). They make no attempt to explain how the observations are transmitted without a medium. Also, MEs, LT and Relativistic Addition of Velocities (RAV) are all medium based, using rectangular axes transforms. Whereas, an oblique axes transform, representing ether-less simulations, resulting in paradoxes and uncertainties, should not have been taken seriously to represent reality. They are non causal mathematical simulations, with no apparent physical or experimental evidence to support them.

25 Einstein's oversight. Einstein's belief in his invariant, ether-less, inertial frame to observe motional properties is not appropriate. Causal predictions require a medium to solve the wave equation and propagate waves. *Propagation defined through the causal solution of the wave equation must always be relative to the medium, not relative to the moving frame, as Einstein believed.* The velocity of light is always with respect to the medium, even when there is system motion relative to the medium. The velocity of light remains invariant in the moving frame only because time and space shrink by exactly the same ratio through motion with respect to the medium. The propagation is relative to the preferred reference frame - the medium, where PTA is created.

26 Ether-less claims. Einstein's assumption, that there was no propagation medium, is not supported by any known data. *Without the medium, PTA is not possible and the EM wave equation cannot be solved. Without a solution, measured events cannot be predicted.* They are non causal, the effect (observed event) could occur before the cause (source event), which cannot happen in reality. Einstein based his ether-less predictions on relative motion between systems without any physical argument or justification. This resulted in his ether-less claims: i) Only relative motion between systems can be detected. ii)

No distinction between a stationary and a constantly moving frame can be made. iii) The medium is redundant. iv) Time travel is possible and v) there is no absolute time and space. All these claims are non causal (false). Einstein's ether-less inertial frame responsible for their beliefs, cannot predict measured observations. It is Lorentz's medium based optical frame that correctly predicts the measured events, using a preferred frame of reference.

27 Medium based SR. However, Einstein's SR does predict measured observations. They are based on a propagation medium. Although Einstein denied the existence of the propagation medium, with his non causal ether-less predictions, it is easily confirmed from inspection of his motional electrodynamics field equations that he used a medium to obtain and solve the wave equation. His field equations use Maxwell's and Lorentz's medium based rectangular axes motional transform, which results in the measured predictions. The same motional properties can be obtained directly from the medium based classical wave equation by simply including Lorentz's time and space contraction through motion, relative to medium, independent of SR.

28 Two aspects of SR. Einstein's SR is therefore not consistent, it has two contradictory aspects. One is concerned with time and space contraction of structures passing through the medium. This, leads to Einstein's famous energy equation, $E=mc^2$ etc., which results directly from the medium based LT. This aspect is not in doubt; it has carried 'relativity' through into the modern age of physics. The second aspect involves rejecting the medium in support of Einstein's relative motion. This automatically makes the essential emission and reception positions and propagation paths indeterminate. It also implies isolated regions of relativity with no continuity and no absolute time and space. This is the ether-less aspect of SR that is not a solution of the wave equation, it is non causal and therefore cannot be measured.

29 Moving media. Einstein's ether-less relativity cannot distinguish between: a) *Propagation in a medium at rest in space.* This is supported by the Cosmic Microwave Background (CMB) radiation, detected by Penzias & Wilson (1965). It is also verified through energy collection increase by motion relative to the CMB radiation, measured by COBE (1992). b) *Propagation in an Earth centred frame* i.e. propagation on Earth to be independent of its orbital speed around the Sun. This is supported by MMX (1887), Sagnac (1913) and Michelson & Gale (1925). Also c) *propagation in a heliocentric frame* i.e. propagation through the Solar System to be independent of its motion through the universe, according to Reasenburg et al (1979). These situations can be explained only if the medium exists, is at rest generally, moves with the Earth locally, and with the Sun and Solar System for inter-solar planetary propagation. Relativists, not believing in a medium, have to make these reference frame transforms without any authority (physical justification).

30 Two types of motion and frame. There are two types of motion. One is where systems move with respect to the medium, causing PTA, including propagation differences between source and observer motion with respect to the medium. The other is where there is no relative motion between the system and medium i.e. no PTA, as in a system and medium at rest in space, or moving with gravitational bodies. Without recognition of the propagation medium, Einstein could not distinguish between these types of motion. Finally, confusion in relativity occurs through there being two types of reference frame: Einstein Inertial Frame (EIF), where the mechanics and the speed of light are invariant, which without a medium is incapable of transmitting observations. And a variant Lorentzian Propagation Frame (LPF) containing the additional effect of the medium and its waves, which make the observations possible, according to Lorentz.

Implications of NR

1 Gravitational Entrainment Model (GEM). A vacuum medium, having no mass (no atomic structure), but having an electrical inertia and stiffness, behaves as an electrical fluid medium. Although it has unity refractive index (cannot convect light), it can refract light through medium compressions and velocity gradients created around moving gravitational bodies. Here the medium moves with the light source and gravitational mass, there is no relative motion. This is not to be confused with convected light where the medium moves relative to the light source. The medium is attracted to, compressed and moves with large gravitational bodies (planets), Schwarzschild (1916)

2 Electro-gravitational Boundary Layer (EGBL). The extent of the 'stationary' medium around planets appears to be controlled by the planet's Gravitational Field of Dominance (GFOD), in the presence of its sun's gravitational field. If the planet rotates, the medium close to its surface tends to rotate with the surface relative to the surrounding 'stationary' medium. This creates an Electro-gravitational Boundary Layer (EGBL) immediately above the surface. The 'stationary' medium surrounding the planet, in its GFOD, then orbits with the planet through the 'stationary' medium in its sun's GFOD and solar system. The medium around the sun and solar system then presumably moves with the galaxy relative to the medium at rest in space.

3 EM Motional Analysis (EMMA). Einstein's medium based theory although causal it is not complete, it cannot distinguish between source and observer motion. A general solution of the wave equation is developed by extending Lorentz's medium based motional theory. This is implemented for source and observer motion relative to the medium, using three sets of time and space scales, source, observer and medium, rather than the two unspecified used in SR. As a result, measured differences between source and observer event times, propagation distances and motional angles are predicted.

4 New space-time. Distinguishing between source and observer motion allows for the first time, both source and observer flight paths to be chartered on the same rectangular axes space-time diagram. *Inter-changing the source and observer for the same flight path gives different observations.* This again negates Einstein's concept of relativity, which cannot distinguish between source and observer motion, claiming only relative motion between systems is detectable. Relativistic time and structural contraction, besides making ageing slower and things heavier, make (i*) propagation time invariant perpendicular to direction of motion* (reflection from carriage window is invariant, but reflections in the direction of motion from a mirror perpendicular to motion is variant), (ii) *Doppler effects for constant source and observer motion identical,* whereas classical effects are different, and (iii*) a moving more dense electrical medium than a vacuum partially convects light.*

5 Time travel. If the speed of light could be exceeded in the medium, or bypassed, perhaps a traveller could go back to the past, see a broken cup reassemble itself, see an exploding supernova imploding, or more macabre, see ancestors resurrecting from their graves. It is possible (causal) to travel visually to the past, but not to interfere or change the past, as it has already happened. It is not possible (non causal) to visually travel to the future as it has not yet occurred. Neither is it possible to materially (physically) travel into the future or past. But it is possible to slow one's time down through material transport, by physically moving at a high speed relative to the medium, or visiting a gravitational body where the medium's time and space is compressed. However, again this is not reversible time travel, it is just relatively changing the rate of ageing between situations.

6 Hybrid frames. Accepting the reality of the propagation medium, and presuming that the speed of light cannot be exceeded in the medium, M (or β)=v/c<1. Where M is the Mach number (1887), system speed 'v' compared to the speed of light 'c' in the medium. It appears that the speed can be exceeded across frames, $M^*>1$. Where $M=v^*/c=v/\alpha c=M/\alpha$. The speed and distance capability is much greater across the hybrid frame by α^{-1}, where α is the Lorentzian contraction factor. Distance is measured in the stationary propagation medium and the slower time in the moving system frame. As $M\rightarrow 1$, $\alpha\rightarrow 0$ and $M^*\rightarrow\infty$, allowing the hybrid speed of light to dramatically exceed the speed of light in the medium, resulting in spectacular distances to be achieved in space travel in a human life time. Space contracts only in the direction of motion, therefore astronauts and the space ship would need to rotate regularly to avoid permanent flattening in one direction, although it could be a painless way of slimming.

7 Gravity and acceleration. The propagation medium is central to both gravity and motion. According to Schwarzschild (1916), gravity attracts and compresses the medium around gravitational bodies (time and space contract). According to Lorentz, motion of systems passing through the propagation medium causes their time and space to contract. Source contraction slows time, observer contraction relatively expands the observed medium and its contents. Time quickens and space effectively expands, reducing gravity and creating a time Equivalence Principle. Gravity, observer acceleration and the LT are all related through an equivalent velocity relative to the medium.

8 Nature of gravity. The only large scale static attractive force known in nature, besides nuclear, is the electrical force between charges. Not only do unlike charges attract, but also groups of unlike charges attract other groups. Van der Waals (1873) established that atoms and molecules attract each other in the near field, through induced dipoles. Gravity can be considered as an extension of this attraction into the far field. The difference electric field between protons and electrons from a large assembly of atoms and molecules constituting gravity. Generally, a finite distribution of positive and negative charges will always produce internally an attractive force and externally an attractive difference electric field.

9 Gravitational field. Although point atoms and molecules with equal numbers of positive and negative charges are neutral, a finite distribution of dissimilar charges (non point) creates an Atomic Residual Difference Field (ARDF), which does not quite cancel, leaving a very weak but finite attractive field. This field has similar properties to gravity, it has an attractive field, it requires a propagation medium, it has a retarded propagation time and an inverse square law decay with distance.

All electric fields appear to use the same electric medium to transmit their steady state values and disturbances. Steady electrical fields are electrostatic, unsteady electrical fields create EM waves, difference electric fields cause gravity, and the residual gravity field from all the gravitational mass in the universe creates the inertial field, similar to that of Higgs (1964). Field disturbances can be represented by waves or non massive particles: light waves as photons, gravity waves as gravitons and inertial field disturbances by bosons.

10 Inertial field. The ARDF from all the gravitational matter in the universe, propagating in the medium, summed over all space, will produce at any point a Universal Gravitational Reference Field (UGRF). This field has zero electrical intensity from the gravitational forces from all directions in space. However, it will create a finite energy field in the propagation medium through expansion against gravitational attraction. The total mass distributed evenly on an expanding annulus ($4\pi r^2 dr$), contributing to any point in space, is proportional to the square of the distance. As the field will decay according to the inverse of the square of the distance, the field will be at least maintained throughout space.

11 Many fields. This attractive 'dark' energy, stored in the UGRF, is available to create the inertia for mass in motion, and though gravitational attraction impede the expanding universe. Whereas, the propagation medium provides a mechanism for possibly creating the medium's repulsive 'dark' energy, through its repulsive expansion, overcoming the gravitational attraction. Thus a vacuum is not empty space, it has its electrical medium with its electrical permeability (inertia) μ and permittivity (stiffness) ę, supporting an attractive gravitational field, and possibly a repulsive medium field. The medium also propagates EM fields (light) and the cosmic microwave background field pervading over all space.

12 Inertial and gravitational mass. As Einstein kept reminding us, energy is equivalent to inertial mass. The inertial mass of subatomic particles, complete atoms and molecules is the binding energy of the system, the energy required to create and hold the structure together. Inertia is the force required to accelerate binding mass (energy) relative to the inertial energy field (UGRF). It appears that it is not the inertia or binding mass m_b, that causes gravity, it has no attractive capability. The attraction appears to come from the ARDF from atoms and molecules having an attractive gravitational mass m_g. Usually $m_g=m_b$, but m_b can exist without m_g, as for example the electron has inertial mass but no bipolar gravitational attraction.

13 Unification. The electrical propagation medium is shown to exist. It provides the common thread that unifies electromagnetism and gravity. The electrical medium transmits steady electric fields, unsteady electric fields (light), steady difference electric fields (gravity) and a universal residual difference field (creating inertia). Gravity is explained through a weak difference electric field from finite distributions of dipoles (dissimilar charges from protons and electrons) within atoms and molecules throughout the universe, explaining gravity in terms of a difference electrical field. The new theory re-establishes a preferred frame of reference and restores the bridge between classical and modern physics. It links Lorentz, accelerating fields and gravity and distinguishes between source and observer motion, providing a basis for the unification theory of the universe.

Chapter I:

A New Relativity theory

1 Introduction

Over a hundred years ago the universe seemed confusing and EM waves not well understood. It appeared that there was no difference between a stationary and a constantly moving system (inertial frame), where the mechanics, the speed of light, Maxwell's Equations (ME's) (1865) and the propagation time perpendicular to the direction of motion did not change through motion. Further, unlike classical systems, there was no difference between receding EM sources and observers (the Doppler effects were identical), and that only relative motion between systems appeared important. Einstein interpreted these observations, as though there was no propagation medium (ether).

Although these observations appear to be correct, they are incomplete. The propagation time is variant in the moving frame, creating Propagation Time Asymmetry (PTA), which invariant inertial frames cannot predict. Also observed time histories are different for the same source and observer flight paths, and motional properties are shown to be dependent on motion relative to the medium. Einstein, evidently was not aware of this additional information. In retrospect, it is physically impossible to create waves without a propagation medium. There is a fundamental wave requirement, inevitably supported by experimental data, both classical and EM, that all wave propagating disturbances require a propagation medium to propagate and make its wave equation predictable (causal). The New Relativity (NR) complies with this requirement. It also removes confusion through distinguishing between ether-less and medium based relativity, which Einstein, not believing in a propagation medium, could not do. Although Einstein denied the existence of the medium, he actually used a medium in his field equations, enabling him to solve the wave equation, making his predictions causal.

In his Special Relativity (SR) (1905), Einstein used Lorentz's medium based motional Transform (LT) (1899) in deriving his motional wave equation, incidentally without acknowledging Lorentz. Einstein's causal theory is therefore based on ME's and the LT, both of which are medium based. Also the Relativistic Addition of Velocities (RAV), essential in combining speeds, is based on the LT and the same medium. The final blow to Einstein's ether-less relativity is that the measured predictions of SR can be derived directly from the medium based classical wave equation, modified by the LT, independent of SR, discrediting the ether-less aspects of SR.

Whilst predicting the same measured observations as the medium based SR and General Relativity (GR) (1915), the New Relativity (NR) accommodates additional measured features, distinguishing between source and observer motion, removing paradoxes, and re-solving long debated challenges in areas where SR and GR have often resulted in lively discussions. Einstein's relativity is only causal, (predictable) when a medium is involved. Those who believe there is nothing wrong with Einstein's relativity cannot answer fundamental questions such as; how does light propagate, or how can one solve the wave equation for propagating waves, without a propagation medium?

Predictions based on Einstein's SR and GR such as Kramer et al (2006) on binary pulsars, are not ether-less, as Einstein would claim. They are accounted for using the medium based NR. Whereas,

Einstein's ether-less aspect of relativity is not measurable, it does not satisfy the medium based causal wave equation. Nor is it supported by the medium based LT, where interaction with the EM medium is essential. Wright (2010)[1]- (2013)[10], shows that ether-less relativity, based on Einstein's inertial frame, is non causal. Its claims and predictions, such as simultaneity, time travel and no absolute time and space, are a simulation not reality. They are based on non causal ether-less oblique transform axes, rather than causal Lorentz's rectangular medium based axes. This causes confusion in terms, such as time dilation when a moving source's time and structure actually contract.

Finally, removing the medium is against basic wave theory creating a discontinuity between classical and modern physics. There is no evidence, or causal model that can support Einstein's ether-less universe. After a century of SR and GR, there have been no significant developments to explain what gravity is and how it is propagated. The new theory identifies and establishes the existence of a well defined measured medium and unifies it with gravity. There is now sufficient evidence to re-think and advance our basic electromagnetic (EM) model, laying the foundations for a new approach, by reinstating the EM medium for the propagation of EM waves and gravity, and further recognising that gravity is electrical. This first chapter is a summary of the main properties of the New Relativity (NR) theory.

2 Medium Aspects

2.1 Lorentz fundamental

It is not generally realised that Lorentz, with help from Poincaré (1900), developed the fundamental EM motional wave theory *based on Maxwell's medium.* Through the Lorentz Transform (LT) the complete measured motional properties are predicted, both classical and relativistic. The prediction is dominated by the motional distortion

(a) Unsteady sources in motion

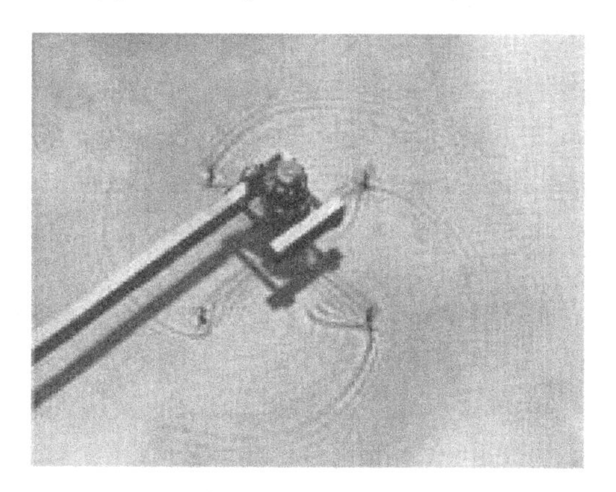

(b) Steady sources exceeding wave speed

Figure 1.1 These Propagation Time Asymmetric (PTA) wave patterns, generated by all forms of source motion, including classical and EM sources, are caused by motion with respect to the propagation medium. There is nothing special regarding EM motional theories. The same basic theory predicts both classical and EM waves at low speeds

of the original disturbance, shown in Figure 1.1. It is characterised by the classical wave Propagation Time Asymmetry (PTA) of the original spherical propagation. It surrounds moving systems, evident in all wave theories. In EM theories, there is the additional modifying Lorentzian time and structural contraction (LC) at high speed. It is possible to have PTA without LC, but not LC without PTA. Classical medium based PTA is therefore the basic characteristic for all motions, absent in Einstein's ether-less theory.

Time and space of the medium do not change, it's the time and structure of physical objects (atoms and molecules) that contract passing through the medium at high speed. Time slows and space shrinks in the direction of motion by the same fraction making the propagation speed of light invariant. *Spacecraft shrink both time and structure in the direction of motion, but not the distances they travel.* LC is not instantaneous, at low speeds it needs to be integrated over a considerable time for its effect to be appreciable. Also at low speeds classical and EM waves become virtually identical, as the same PTA dominates both types of motional effects. So in spite of Einstein's speciality claim, there is nothing special regarding EM waves, they require a medium the same as all causal wave theories. Notionally, the LT can be expressed by the incremental equation:

$$\Delta LT=(\Delta PTA)_{classical}+(\Delta LC)_{relativistic} \qquad (1.1)$$

The above expression is determined through the solution of its wave equation, using a medium with finite properties, the same as any other wave propagation theory. There is no other rational outcome, Lorentz predicts no ether-less properties. He predicts measured events according to the system's instantaneous motional interaction with the medium. The theory is causal, the cause (source event) must always occur before the effect (observed event), which is not the case for the ether-less aspect of Einstein's theory.

2.2 Propagation time asymmetry

PTA is the result of a preferred propagation frame, it is of course variant, not invariant according to Einstein's inertial frame. It determines the reality of wave propagation that everyone is familiar with, creating wave asymmetry around a moving system. It's irrational to accept that EM sources are somehow unique, that they lose their propagation medium when set in motion. *The conventional Galilean concept of PTA was extended by Lorentz, his Transform (LT) includes PTA as an integral part.* It provides the foundation for Einstein's medium based causal SR. Einstein's ether-less relativity, without a medium, cannot predict PTA. For M<<1, the classical PTA part of equation 1.1, through motion with respect to the propagation medium, is predicted by:

$$\Delta t=\pm\Delta d/c=\pm vt/c=\pm(vd/c)/c=\pm vd/c^2=\pm Mt, \qquad (1.2)$$
$$M=v/c,\ \ t=d/c,\ \ \Delta t/t=\Delta d/d=M$$

Where v is the system velocity, ± is for motion in and against the direction of the wave propagation, c is the propagation speed and d is the distance between two fixed points moving with the system relative to the medium. Figure 1.1(a) is for a single frequency and system Mach (1887) number M (or β)=v/c=0.66, which determines the degree of distortion (66%) from spherical symmetry. Figure 1.1(b) is a wave pattern generated by the PTA part of the Lorentz transform, LC=0 in equation 1.1, for an electron moving for M>1. Again, the classical medium based PTA predicts the main structure for an impulsive wave front from an electron exceeding its wave speed in the medium, as in Cerenkov (1934) radiation. Here a high speed electron leaves a vacuum and enters an electrically more dense medium, with a lower wave speed (air).

2.3 Einstein's inertial frame

Einstein's Inertial Frame (EIF), upon which ether-less relativity is based, provides no answer on how light and gravity are transmitted across the cosmos and cannot identify any authority allowing the order between cause and effect to be reversed. It is correct that Einstein's constant motion inertial frame, where the mechanics and speed of light are invariant, is non preferred (not unique). However, Einstein's ether-less frame cannot predict observed motion as he believed. *To observe the events requires the propagation medium to be restored to create PTA and allow the events to reach the observer.*

Thus the ether-less, non preferred, invariant, omni-directional inertial frame, requires the wave propagation to be included. It then becomes the medium based, preferred, variant, directional, propagation frame, which propagates the source events to the far field. Einstein it seems, misunderstood the basic wave propagation process. Attempting to eliminate the medium is irrational, it violates well established physical principles, preventing a causal solution of the wave equation, and removing the propagating means for waves that produce the observed effect (transmission of energy and information). There is no justification for removing the medium. It appears that *Lorentz's medium based motional theory, extended by NR's wave equation solution is the legitimate and complete description of wave propagation with respect to the reference medium.*

Through extensive measurements, medium propagation is shown to be the case, there is no other causal description. If a system moves relative to the medium it does not affect the wave propagation speed in the medium, according to Lorentz's basic wave theory. But it does result in PTA and variance in the moving frame. Einstein's Inertial Frame (EIF), where it is assumed there is absolutely no difference between a stationary and constantly moving system i.e. the mechanics, propagation speed and propagation time are all considered invariant, cannot provide a solution to the wave equation. *A frame with no medium cannot predict observations, it is non causal, it is purely an internal moving frame description, which cannot account for external observed wave propagation.*

2.4 Einstein's oversight

Although the propagation speed and mechanics are invariant in a constantly moving frame, constituting Einstein's two postulates, to observe these effects a third postulate is required: a medium is needed to transmit the observations and accommodate PTA. According to Causality there are two conditions. The medium can:

I. Either move with the system, where there will be no PTA ahead and behind the moving system.

II. Or the systems can move relative to the medium, where there will be PTA.

But propagation defined through the causal solution of the wave equation, must always be relative to the propagation medium, not relative to the moving frame as Einstein believed. This is Einstein's basic error; wave propagation must always be relative to the preferred reference frame - the propagation medium. The moving frame is not a propagation frame. Its propagation speed remains invariant only because both time and space contract by exactly the same ratio through motion with respect to the medium. This makes the propagation speed invariant, but not the measured PTA laid down in the stationary medium. It is Lorentz's medium based theory, extended by NR's general wave equation, based on a medium that appears to correctly predict the variant motional properties for both source and observer motion. Einstein's invariant inertial frame, responsible for predicting the ether-less properties, is incapable of supporting observed reality.

2.5 Motional effects are classical

Not only are the EM disturbances medium based, but relativistic arguments cannot explain them. The Lorentzian contraction, created passing through the medium, and the Schwarzschild (1916) gravitation compression of the medium, discussed briefly in Section 7, are small at Earth speeds and gravitational strengths. Whereas, the classical PTA is the fundamental instantaneous disturbance. It is shown in Section 3 that the basic EM motional effects: Michelson and Morley (1887), Sagnac (1913), Michelson and Gale (1925), Saburi et al (1976) and GPS (1992), cited supporting the relativistic effect at these low speeds, are in fact explained using only the classical PTA wave theory, the first part of equation 1.1. *If this variant PTA effect for EM systems in motion had been accepted in 1905, it could have prevented Einstein's non causal ether-less, invariant aspect of relativity from developing.*

On Earth, motional effects, without involving particle accelerators or particles from outer space, are predicted by the medium based classical wave theory, again just the classical PTA part of equation 1.1, i.e. equation 1.2. Relativistic and gravitational contributions have little effect on the instantaneous PTA disturbance. Assuming realistic Earth source speed of $M_s \approx 10^{-6}$ and Earth equivalent gravitational speed of $M_g \approx 10^{-9}$. From equations 1.2, 1.9 and 1.12, we have for the time ratios, compared to the unaffected values, Lorentz $\Delta\tau_s/\Delta\tau_p = -M_s^2/2 \approx -10^{-12}/2$, gravity $\Delta\tau_g/\Delta\tau_p = -M_g/2 \approx -10^{-9}/2$ and PTA $\Delta t_{pta}/t_p = M_s \approx 10^{-6}$. *It is the classical PTA wave disturbance that is the basic feature.* These effects are multiplicative, where relativistic and gravitational effects are a small modifying effect of the PTA at speeds and gravitational strengths available on and around the Earth, and small integration times.

2.6 LT and SR

The Lorentz Transform (LT) is fundamental; it is not the same as Special Relativity (SR), which has two aspects. One aspect of SR claims no propagation medium, predicting ether-less observations that are non causal and cannot be measured. The other aspect predicts measured causal observations through its field equations based on a medium (LT). When SR is claimed to have been verified, it is actually the LT or medium based SR version that is usually measured, not the ether-less non causal aspect. Without robust reasoning, Einstein put his faith in his ether-less invariant inertial frame to attempt to support the invariant Michelson and Morley Experiment (MMX) (1887). Without a medium, ether-less SR cannot detect constant motion or predict PTA. Whereas, the variant medium based PTA, is the vital part of any causal wave theory, both classical and EM. Einstein's inexplicable concept of ether-less relative motion, denying PTA, offers no alternative mechanism to that provided by the medium.

Einstein's belief that his SR and GR needed no medium to propagate EM waves and gravity is therefore in error. Einstein's measured predictions are shown to be based on a medium, as Lorentz predicted, not on relative motion between systems as Einstein claimed. *All EM motional observations can be predicted directly (without Einstein's SR) by simply using the medium based classical wave equation. Modified additionally, for high speed motion, by the medium based LT contraction.* Therefore, by far the major observed motional effect, on and around the Earth, is the classical PTA. It is validated every day, it accounts for aircraft inertial guidance systems and in the workings of Global Positioning Systems (GPS).

3 Re-interpreted Experiments

Einstein's ether-less relativity claim was based mainly on the null effect on light propagation on Earth as it moves through space, according to the Michelson and Morley Experiment (MMX) (1887) shown in Figure 1.2. Einstein believed that the MMX supported his inertial frame, where there is no difference between a moving and stationary frame. This is not possible, according to causal wave theory, Condition I, Section 2.4. Here waves without PTA must mean that there is no relative motion between the system and medium.

The MMX is explained quite naturally with the medium moving with the Earth. Even if there was relative motion, relativistic effects could not explain the null result. It's true, at right angles to motion the extra

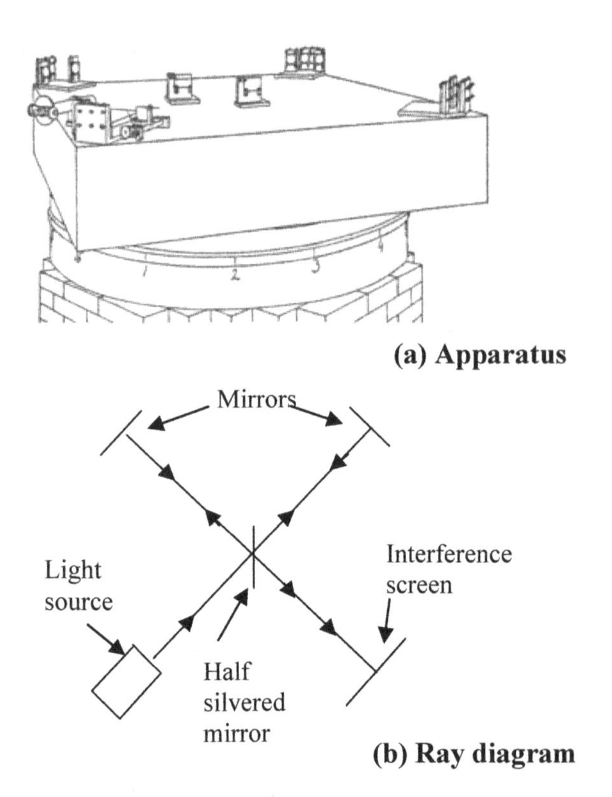

Figure 1.2. This is the famous Michelson and Morley Experimental (MMX) (1887). Figure (a) shows the apparatus. It involves measuring the light propagation in the direction of the Earth's motion and at right angles, using a monochromatic light source, mirrors and an interference screen. This system can measure propagation time asymmetry (PTA) between the two directions by rotating the apparatus. The PTA is measured, through interference band movement on the screen, shown in Figure (b). No band movement (no PTA) was measured, although the Earth was travelling at considerable speed ($M=10^{-3}$) through space. It was believed that the null result indicated that there was no optical difference between a stationary and constantly moving system (inertial frame), implying there was no propagation medium (ether). As a result Einstein developed his ether-less aspect of relativity, which has never been measured. It is obvious, in hindsight, that the ether-less concept is in error. The MMX null result satisfies causality (Condition I), whereas the ether-less concept does not. The medium does in fact exist and moves with the Earth, as illustrated in the Gravitational Entrainment Model, Figure 2.8, Chapter II

distance the light travels in the medium is offset by the time slowing in the moving frame making no PTA. However, in the direction of motion, the effect is small, and the variant PTA dominates, as explained in Section 3.3, Chapter IV. If there had been relative motion in the direction of motion, the MMX would not have been a very sensitive method of measuring motion, as discussed in Section 6.2.

The absolute demonstration of the medium's presence was performed as early as 1913 in Sagnac's classical rotating mirrors experiment. Sagnac passed light through the medium around a loop via four rotating mirrors in a square of perimeter d, as illustrated in Figure 1.3(a). The time t it takes for the light to pass around the loop in the medium is $t=d/c$. The small distance the loop rotates in the medium in this time at speed v, is $\Delta d=vt=vd/c$. The time for the light to travel this distance is $\Delta t=\Delta d/c =vt/c=Mt$, which is the classical PTA part of equation 1.1, i.e. Equation 1.2. The measurement was predicted exactly according to this equation confirming the presence of the medium.

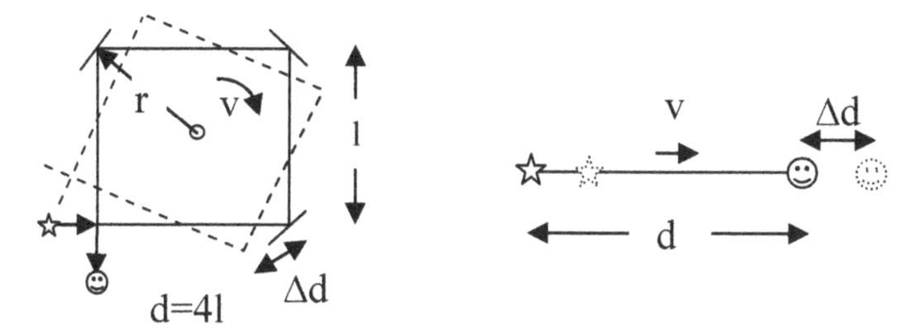

(a) Sagnac's rotating mirrors **(b) Equivalent linear system**

Figure 1.3 In Figure (a) Sagnac's rotating mirrors experiment generates a light propagation delay Δd. The delay depends on the speed and direction of the mirrors relative to the propagation medium. In Figure (b) the circular motion is reconfigured as an equivalent linear motion system. Sagnac demonstrated motion exactly according to equation 1.2, establishing the medium's presence

Figure 1.3(b) is the equivalent linear motion configuration. These two experiments, the MMX and Sagnac's investigation, are based entirely on classical physics with medium propagation. Claims that these experiments can be accounted for through relativistic effects are false, effects are negligible compared to PTA, as discussed in Section 2.5.

The acceptance of any new theory is by measurement confirmation. The NR theory is upheld through all well known data. *Michelson and Morley Experiment (MMX) (1887), Sagnac (1913) and Michelson and Gale (M&G) (1925), sometimes thought to support the lack of a medium, are all explained through the medium based classical PTA. As* described in the medium based Gravitational Entrainment Model (GEM), Section 6, Figure 2.8, Chapter II. The MMX has no PTA, there is no motional effect on light propagation in the stationary medium rotating with the Earth. This is simply because the measuring equipment is stationary relative to the medium. Sagnac, where his mirror motion is in and against the light propagation, relative to the stationary medium on the Earth's surface, predicts PTA exactly.

Further M&G, demonstrates that the medium on Earth clings to the Earth's surface, increasing its speed towards the equator, relative to the surrounding medium. This indicates that there is an Electromagnetic-gravitational Boundary layer (EGBL) between the rotating Earth surface and the surrounding stationary medium. The early interpretation of this experiment was that if the medium existed and rotated with the Earth out to infinity, there would be no relative motion. The existence of the EGBL illustrated in Figure 2.8, now invalidates this objection, supporting the medium's presence. M&G's investigation is described in detail in Section 5.3, Chapter IV.

Similar supportive arguments apply to more modern data, Saburi et al (1976), Reasenberg et al (1979) and Global Positioning Systems (GPS) (1992). Saburi (satellite communication across the Pacific) caused measured PTA of 300ns (100m) in and against the Earth and satellite's rotation with respect to the surrounding stationary medium, according to equation 1.2, described further in Section 3.7, Chapter VIII. Reasenberg et al, established that the medium moved with the

Sun and Solar System at high speed ($M=10^{-3}$) through the universe. They established in Mars-Earth communications that there was only local motion within the medium, i.e. negligible PTA.

If the medium was not moving with the Solar System then there would have been a large PTA transmitting across the Solar System, relative to the stationary medium in space. Also in GPS, where satellite and Earth stations rotate relative to stationary medium surrounding the Earth, a one way PTA is created, causing a predicted measured surface positional displacement of 30m, using equation 1.2, discussed further in Section 3.5, Chapter VIII. For satellites, there is need for minor adjustments which results from the long term drift associated with relativistic and gravitational effects and passing through non homogenous media such as the ionosphere. Further details of the above references are considered in Section 6, Chapter II.

PTA is the tell-tale evidence for causal predictions when systems move relative to the medium. The examples above, based on a propagation medium, are just a few of many that show Einstein's ether-less, relative motion interpretation is in error. *All six of these basic experiments are accounted for, through the classical PTA using a propagation medium.* Claims that relativistic effects account for these motional changes are not possible at Earth speeds and short integration times. Inspection of Einstein's field equations in his SR show that in fact he used a propagation medium, and further did not comply with his own relative motion claim. *Thus Einstein believed there was no medium but used one in his field equations.* This caused confusion, it gave Einstein the capability of predicting causal (measured) properties, including PTA and LC inherent in LT, according to equation 1.1, rather than the ether-less predictions he claimed.

Therefore, Einstein's SR has a medium based causal aspect that predicts many of today's measured observations, but is incomplete. And an ether-less, non causal aspect, which predicts ether-less properties that cannot be measured. NR is a complete, robust all-inclusive, theory that can provide simple solutions for overly complicated approaches to problems. For example, the above references can now be basically understood using simple school level physics. Relativistic effects claimed to interpret these types of experiments, are not significant, they only become important over long measurement periods, or high relativistic speeds.

4 Medium's Credibility

Electromagnetic (EM) waves and gravity, which are supposed to propagate without a propagation medium (ether), by definition are a contradiction, and in reality physically impossible. In the absence of all gravitational matter, including fields and particles, a vacuum is not empty space. It has long been established that a vacuum is filled with an electrical medium having measureable properties, (electrical permeability, inductance (inertia) $\mu=1.25\text{x}10^{-6}$ N/A^2 and electrical permittivity, capacitance (stiffness) ę (or ε) = $8.85\text{x}10^{-12}$ F/m), which enables waves to 'bounce' through the vacuum. With these well defined properties, there is little point in denying the medium's existence. It interacts with moving systems (atoms and molecules) contracting the system's structural time and space according to Lorentz and is compressed by gravity according to Schwarzschild (1916).

The medium also determines the wave characteristics, including the wave propagation speed given by $c=(\mu ę)^{-1/2}$. The electrical medium properties being finite give a finite speed of light. *If there was no medium $\mu ę=0$, there would be no PTA or Doppler effect around a moving system and the propagation speed in space would be infinite, which is not the case.* Simultaneity (equal propagation times upstream and down) and reciprocity (interchanging the observer and source makes no difference optically), are claims of an ether-less relativity, that cannot be measured. *Simultaneity and reciprocity cannot be* *supported by the medium based Maxwell's Equations (ME's) (1865), and the medium based Lorentz Transform (LT) (1899). These claims are avoided by simply accepting the medium's presence.*

Motion relative to the medium is readily verified, either by inspection of Einstein's field equations that clearly use a medium in Lorentz's medium based transform. Or absolutely derived directly from first principles, where the medium based classical wave equation is modified by Lorentz's time and space contraction through motion. Either way, Einstein's measured SR observations are predicted using a medium. Relativists interpret Minkowski's (1908) rectangular axes space-time four vector analysis as requiring no medium. However, this is not possible; the vertical and horizontal axes of space-time still require Lorentz's rectangular axes transform, representing time and space, respectively. Normalised against the speed of light, the velocity becomes a 45 degree gradient. This is just a mathematical convenience of representing time and space; *there is no mechanism or justification for removing the propagation medium.*

Thus all wave theories require a medium to be rational. The medium is not a mathematical artefact that can be removed, as Einstein believed. *The medium is needed to support the field that supports the waves that illuminates the observed events.* The medium's presence should not have been doubted over a hundred years ago, there was considerable evidence then to support the medium. There is no justification now to reject the medium, and *there has never been any evidence to support Einstein's ether-less aspect of relativity.* All motional effects, both classical and relativistic, are shown to be caused by system motion relative to the medium, as Lorentz predicted. Classical and EM systems behave similarly, to make waves and transmit information, both require a medium.

The confusion in Einstein's relativity is evident through Einstein not being able to distinguish between two types of reference frames.

Einstein believed that EM waves were somehow different from classical ones. *It was acknowledged that the medium existed for stationary EM systems, but it is irrational to suddenly disappear when the systems were set in motion.* This inconsistency was suggested by Einstein's ether-less inertial frame, but without a propagation medium it has no capability of propagating and predicting observations. The measured situation is described by Lorentz's medium based variant optical propagation frame, which makes the judgment of observations possible. *PTA is ensured, it is a classical and causal prediction, passing through into Lorentz's medium based transform, comprehensively invalidating Einstein's ether-less concept of relativity.*

Further, confusion is apparent in Einstein's relativity through not being able to distinguish between various types of frame motion. One is where a system is stationary or moves at constant speed. Another is between source and observer motion. A third is between stationary and moving medium frames, as in distinguishing between propagation in the medium at rest in space, moving with the Earth (Earth centred), confirmed by the MMX (1887), and moving with the Sun and Solar System (Heliocentric), established by Reasenberg et al (1979). *These frames are implemented without authority; they can be explained only if the medium exists, is generally at rest, moves with the Earth and finally moves with the Solar System,* supporting the new Gravitational Entrainment Model (GEM) discussed in Section 6, Chapter II.

Einstein's ether-less predictions were supported by Born (1924), described in Section 3, Chapter III. He adopted Minkowsky's oblique transform axes representation, rather than the well established rectangular axes used in the Lorentz transform. However, these predictions are simulations based on no medium. They attempt to remove the effect of the medium by using medium based concepts (circular argument). Without a medium the predictions cannot be a solution of the wave equation; they are non causal and cannot be measured. Thus Einstein's claims: i) only relative motion between systems can be detected, ii) distinction cannot be made between frames moving at various constant speeds and iii) the medium is redundant, are all non causal. *Einstein's inertial frame, responsible for these beliefs, without a medium, cannot predict measured SR observations. It is the medium based LT that predicts them.*

5 New Relativity

To argue that the medium does not exist has caused uncertainty in the understanding of EM wave theory and a tendency towards metaphysical speculation. It has led to the discontinuity between classical and modern physics, prevented attempts to unify EM and gravitational fields, and possibly the ultimate unification theory of the universe. Relativists not accepting the medium's existence are reluctant to accept that systems contract, both time and structure, passing through the medium, according to Lorentz (1899), equation 1.9, Section 7. Also that gravity attracts and compresses the medium, both time and space, (space-time distortion), according to Schwarzschild (1916), equation 1.12. *Basic optics tells us that all fundamental wave concepts, such as Huygens's (1678) wave theory, ME's and LT are all medium based. They are meaningless without their medium to propagate their steady fields and disturbances.*

This includes the propagation speed of light, where the light propagates relative to the medium. The speed remains invariant in the moving frame through time and structure of the moving system contracting by exactly the same ratio. Maxwell's Equations (ME's) are also developed for wave motion within the propagation medium. They are invariant in a moving frame because their final differential form depends only on the speed of light, which is invariant in the moving frame. Although the physics and mechanics (Einstein's first postulate) and the speed of light (second postulate) are invariant and correctly represented in Einstein's inertial frame, the propagation time is considered invariant, which is not what is actually measured. The propagation time is always found to be asymmetrical and variant around a moving system. *Therefore, in the New Relativity (NR), a third postulate appears to be required stating that a propagation medium is essential to create Propagation Time Asymmetry (PTA) and its variance around a system in motion.*

A causal EM Motional Analysis (EMMA), is described briefly in Section 7, Chapter II and in detail in Chapter V. It extends the medium based Lorentzian transform to include both source and observer motion, clearly evident in the measured data. With no medium, Einstein believed no distinction could be made between these two motions. *Accepting the mediums presence now requires distinction between the source and observer motion.* Both the source and observer's time and structure will contract through motion relative to the medium. *However, time slowing and structural contraction at the observer relatively causes the observed surroundings, including any medium compression through gravity to effectively expand,* with the capability of neutralizing the effects of gravity.

Therefore, NR with its propagation medium creates links between LT contraction through source motion, medium compression through gravity, and accelerating frame expansion of the medium through observer acceleration. This is the optical or time description of the Equivalence Principle. EMMA distinguishes between measured motional times and spaces compared to the stationary medium. It uses three individual pairs of time and space scales representing source, observer and medium, illustrated in Figure 2.9, rather than the two unspecified pairs in the incomplete SR. *NR predicts measured properties that ether-less SR cannot predict,* including different source and observer motional wave propagation angles, source motion contracting time and space, observer motion expanding them, and different propagation times, observations and motional paths.

After a century of adjusting to an ether-less universe and the possibility of time travel, NR confirms that it is only possible (causal) to travel visually to the past, but not to interfere or change it, as it has already happened, Section 3, Chapter V. If the speed of light could be exceeded in the medium, or bypassed, perhaps a traveller could go back to the past, see a broken cup reassemble itself, see an exploding supernova imploding, or more macabre, see ancestors resurrecting from their graves. However, *it is not possible (non causal) to visually travel to the future, by definition, it has not yet occurred. Neither is it possible to materially (physically) travel into the future or past, it is against all known physical principles.* But it is possible to slow one's time down through material transport, by physically moving at a high speed relative to the medium, or visiting compressed medium surrounding large gravitational bodies. However, this is not reversible time travel; it's changing only the rate of ageing between systems. One can still go travelling and return younger than one's stay at home family.

Accepting the reality of the medium one can now compare moving frames with the stationary medium. Here it is presumed that the speed of light cannot be exceeded in the medium, M (or β)=v/c<1. Where M again is the Mach (1887) number, system speed 'v' compared to the light speed 'c' in the medium). However, it appears that the speed of light can be exceeded across frames ($M^*>1$, where $M^*=M/\alpha$), described in Section 3, Chapter V. The speed and distance capability is much greater across the hybrid frame by α^{-1}, where α is the Lorentzian contraction factor. Distance is measured in the familiar stationary propagation medium and the slower time and local spatial contraction in the moving system frame. *As $M\rightarrow 1$, $\alpha\rightarrow 0$ and $M^*\rightarrow\infty$, allowing the hybrid system speed to exceed the speed of light in the medium in a spectacular way. Space craft shrink, both time and structure, but not the distances they travel.*

This results in almost unlimited distances to be achieved in space travel in a human life time. Hybrid speed is not mathematical smoke and mirrors, it results in a real practical reduction in space travel time. *It appears that all electrical fields use the same medium to transmit their steady state values and disturbances.* It seems that unsteady electrical fields create EM waves, steady difference electric fields create gravity, and net gravity fields throughout the universe create a residual inertial field. Field disturbances can be represented by waves or non massive particles: light waves as photons, gravity waves as gravitons and inertial field disturbances by bosons. They all appear to be electrical, requiring the same electrical medium and propagating at the same speed.

There is ample evidence to support the medium, but no evidence to support its non existence. The medium based NR provides a link between LT, SR, GR, PTA, accelerating frames and the Equivalence Principle. NR removes the non causal predictions, removing the confusion surrounding relativity. *NR establishes reality from ether-less beliefs, restores the connection between classical and modern physics and provides a possible link to the theory of everything.* Lorentz is fundamental, Einstein's ether-less aspect of SR is a misinterpretation of Lorentz and NR is an extension of Lorentz's motional theory for both source and observer motion. Although Einstein's SR has served well for the last 100 years, NR shows that LT is fundamental, SR is incomplete and its ether-less interpretation non causal.

6 Experimental Verification

One of the main reasons why the propagation medium is not readily detectable, or acceptable, is that there is no existing dedicated measuring system sensitive enough to detect it. The Michelson and Morley Experiment (MMX) (1887) technique, described in Section 2, and Section 6.2, is not a sensitive method of measuring motion with respect to the medium, even if the medium had been at rest in space, not moving with the Earth. At Earth orbital speeds of 30 km/s, or

Mach number $M=v/c=10^{-4}$, only a small fraction of an interference fringe (2π radians) could have been measured, and at Earth's rotational speed of 480 m/s, $M \approx 1.5 \times 10^{-6}$ nothing at all would have been detected.

To establish the medium's presence and measure motion with respect to the medium at practical speeds on Earth, a more sensitive measuring system is required. The MMX is insensitive because it is based on a second order velocity effect (M^2), see Section 6.2, equation 1.5. It relies on the difference in propagation times in each direction in a round trip propagation measurement, where the differences in each direction tend to cancel. The sensitivity can be increased by many orders of magnitude (10^6), compared with the MMX, by measuring the propagation time in just one direction only, or opposite directions separately, as discussed further in Section 6.1.

For those who wish to establish the details of the propagation medium, several methods are suggested. Firstly, the Electro-gravitational Boundary Layer (EGBL) profile above the Earth's surface, described in Section 6, Chapter II, needs to be measured in detail. Here a medium boundary layer is clinging to the Earth's surface according to the MMX (1887), Section 3. Also supported by Sagnac (1913) and Michelson and Gale (1925), as described in Section 5.1 Figure 4.4, and Section 5.3 Figure 4.5, respectively, Chapter IV.

The medium above the EGBL is stationary surrounding and orbiting with the Earth. This is according to GPS (1992), Section 3.5 Figure 8.5 Chapter VIII, and satellite communication across the Pacific, Saburi et al (1976) Section 3.7 Figure 8.6 Chapter VIII. Also Hafele and Keating (1972), discussed in Section 5.7 Figure 4.6 Chapter IV used atomic clocks and flew at 10km above the Earth, in the medium surrounding and orbiting with the Earth. They showed symmetrical time slowing flying in and against Earth's orbital motion (not its rotation where there is asymmetry). This confirms the medium at this altitude is 'stationary', not rotating with the Earth.

The EGBL medium velocity profile above the Earth could be measured at various heights using a tethered balloon, or an aircraft. Motion could be measured using a sensitive instantaneous Medium Velocity Measuring System (MVMS) described briefly below. A bigger task would then be to map the medium velocity profiles around planets, the Sun and moons in the Solar System and elsewhere, using orbital satellites and space craft. *But first it would be appropriate to establish the medium's presence. This could be readily achieved by measuring motion relative to the medium at rest on the Earth's surface, using the MVMS in a train or well sprung vehicle.* Atomic clocks could be used to measure velocity relative to the medium through time contraction through motion. However, the problem with atomic clocks is that they need a long measurement time and are also sensitive to time compression variations through gravity.

6.1 Medium Velocity Measuring System (MVMS)

The sensitivity of measuring motion with respect to the medium (cosmic speedometer) can be increased many orders of magnitude compared with the MMX, by measuring the propagation time in one direction to the motion, or opposite directions separately. An apparatus based on the above concept could measure motion with respect to the medium, or measure the medium's motion relative to the propagation medium, generally at rest in the universe. This cosmic or Inter-Galactic Speedometer (IGSO) could now allow the velocity of spacecraft to be determined, light years from any local object. It is also well capable of measuring an object's velocity with respect to the ground (relative to the medium stationary with the Earth's surface).

Further details of the measuring system cannot be discussed because of a pending patent application. However, this system is well capable of measuring typical velocities on Earth, as well as the higher velocities in space. The propagation time t between two fixed points of path length d, is $t=d/c$. If the system moves at velocity v, then the incremental distance travelled by the system in time t is $\Delta d=vt=vd/c$. The incremental propagation time Δt and phase change $\Delta\varphi$ are then:

$$\Delta t=\Delta d/c=vd/c^2=Mt, \text{ and } \Delta\varphi=2\pi f\Delta t, \text{ where } M=v/c \text{ and } t=d/c \quad (1.3)$$

If N is the number of interference band (dark and bright fringes, 2π phase rotations), f is the frequency and λ the wavelength of the source (laser), then:

$$N=\Delta\varphi/2\pi= f\Delta t=fvd/c^2=fMd/c=Md/\lambda \quad (1.4)$$

If v=300m/s (670miles/hour), $M=10^{-6}$, $\lambda=6x10^{-7}$m and d=3m, then $N=10^{-6}x3x6^{-1}x10^{7}=5$ fringes, which is easily measured.

6.2 MMX Velocity Measuring System

On the other hand, the sensitivity for the Michelson and Morley Experiment (MMX), which measures the Round Trip Propagation Time (RTPT) from Section 4.2, equation 6.11, Chapter VI, is:

$$t=dc^{-1}\alpha[\{1/(1-M)\}+\{1/(1+M)\}]=dc^{-1}\alpha 2/\alpha^2=\alpha^{-1}2d/c\approx(1+M^2/2)(2d/c)$$

$$\text{giving } \Delta t\approx(M^2d/c) \quad (1.5)$$

$$\text{as } [\{1/(1-M)\}+\{1/(1+M)\}]=2/\alpha^2 \text{ and } \alpha^2=(1-M)(1+M)=(1-M^2) \quad (1.6)$$

$$\text{then } N=f\Delta t=fM^2d/c=M^2d/\lambda=10^{-12}x3x6^{-1}x10^{7}=5x10^{-6} \text{ fringes} \quad (1.7)$$

This is one million times less sensitive than the MVMS capability, making the MMX technique impractical to measure motion.

7 Lorentz, Gravity and Acceleration

Now that gravity (steady difference electric) and light (unsteady electric) use the same electric field, sharing the same propagation medium, a direct link can be made between the Lorentz transform, gravity and accelerating frames. Sources, according to Lorentz, moving through the medium contract their time and structure. The medium, according to Schwarzschild, is compressed both time and space, by gravity. An accelerating observer contracts its time and structure, relatively expanding the observed surrounding medium, with the ability to neutralise gravity. PTA depends on the propagation time between a fixed source and observer and the speed the system moves relative to medium.

7.1 Constant motion

Concentrating now on the permanent time changes and ignoring temporary Doppler ones. From Section 4, equation 5.28, Chapter V, for $\varepsilon_s=\varepsilon_o=1$, and a stationary observer $\alpha_o=1$, $\tau_p=\tau_o$, then a source moving at constant velocity 'v_s' and interacting with the medium, contracts its structural time τ_s and space x_s, compared to the propagation medium values τ_p and x_p, thus from equation 5.18 and 5.30:

$$\tau_s= \alpha_s\tau_p, \quad x_s= \alpha_s x_p, \quad \alpha_s=(1\text{-}M_s^2)^{1/2}, \quad M_s=v_s/c \qquad (1.8)$$

M_s (or β_s) is the system (source) Mach number. For low speeds $M_s<<1$, incremental changes are then:

$$\Delta\tau_s/\Delta\tau_p=\Delta\alpha_s=-M_s^2/2, \quad \Delta x_s/\Delta x_p=\Delta\alpha_s=-M_s^2/2 \qquad (1.9)$$

For example, the amount of time slowing for a system moving for 24 hours relative to the stationary medium on Earth, at v_s=300 m/s, (670 miles/hour), $c=3x10^8$, $M_s=10^{-6}$, time slowing (24 hours, $8.6x10^4$s) $\Delta\tau_s$= $-(10^{-12}/2)\Delta\tau_p$ $=-(10^{-12}/2)x8.6x10^4$ =-43ns/day. Time slowing at Earth's rotational speed at the equator (460 m/s) is -95ns/day relative to the surrounding stationary medium.

Flying supersonically (460m/s, $M=1.5x10^{-6}$) in the direction of the Earth's rotation (920 m/s relative to surrounding stationary medium) is from equation 4.17, -4x95=-380ns/day and relative to the Earth's surface, equation 4.16, -3x95=-285ns/day. Flying against the Earth's rotation, 0m/s relative to the surrounding stationary medium is 0s/day, and +95ns/day faster relative to the Earth's equator time. For a moving observer:

$$\Delta\tau_p/\Delta\tau_o=M_o^2/2, \quad \Delta x_p/\Delta x_0=M_o^2/2 \qquad (1.10)$$

As the moving observer's time and space shrink, the surrounding medium and source's time and space relatively expand, (M_o positive).

7.2 Gravitational acceleration

From Section 7, Chapter V, according to Schwarzschild (1916), the propagation medium and its contents in a gravitational field 'g' are compressed both time τ_g and space x_g, compared to their propagation free-field values τ_p and x_p, thus from equation 5.34:

$$\tau_g= \delta_g \tau_p, \quad x_g= \delta_g x_p, \quad \delta_g=(1-M_g)^{1/2}, \quad M_g=v_g/c=2Gm^*/c^2R_g=2gR_g/c \qquad (1.11)$$

$$\text{as } g=Gm^*/R_g^2 \text{ and } v_g=2gR_g/c$$

For $M_g<<1$, incremental changes are:

$$\Delta\tau_g/\Delta\tau_p =\Delta\delta_g=-M_g/2, \quad \Delta x_g/\Delta x_p=\Delta\delta_g=-M_g/2 \qquad (1.12)$$

For example, the amount of time slowing on the Earth's surface due to the Earth's gravity compared with free space, where $G=6.67x10^{-11}$ Nm^2/kg^2 is the gravitational constant, $m^*=6x10^{24}kg$ is the gravitational mass of the Earth, $c=3x10^8$ m/s, $g=9.8m/s^2$, $G/c^2=7.4x10^{-28}$, $R_g=6.4x10^6$ is the radial distance from the Earth's centre, gives $M_g\approx 2gR_g/c^2$

$=1.3x10^{-9}$ for the Earth's surface equivalent gravitational Mach number. For an accumulative time (24 hours) $\Delta\tau_g = -(M_g/2)\Delta\tau_p = -(1.3x10^{-9}/2)(8.6.10^4) = -56\mu s/day$

7.3 Observer acceleration

For an accelerating observer a_o, over a time t_o, and distance d_o and wave propagation distance d_w, from equation 5.38 we have:

$$\Delta\tau_o/\Delta\tau_p = \Delta\delta_o = M_o, \quad M_o = v_o/c = a_o t_w/c = a_o d_w/c^2, \text{ where } v_o = a_o t_w, \; t_w = d_w/c, \quad (1.13)$$

$$d_o = a t_w^2/2$$

For example, the special case of an observer falling under gravity. M_o is positive, observer's surrounding time relatively expands, neutralizing gravity's time loss. According to equation 1.12, $M_o = M_g/2 = 1.3x10^{-9}/2$, time quickens $\Delta\tau_o = M_o\Delta\tau_p = (1.3x10^{-}/2)(8.6x10^4) = 56\mu s/day$ neutralizing the Earth's gravity time slowing. Also $a_o = 9.8m/s^2$, wave distance $d_w = M_o c^2/a_o = 6x10^6 m$, $t_w = d_w/c = 2x10^{-2}s$, observer velocity $v_o = a_o t_w = 2x10^{-1}$ m/s and observer distance $d_o = a t_w^2/2 = 10x4x10^{-2}/2 = 0.2m$

7.4 Propagation time asymmetry

From equation 1.2, $\Delta t_{pta} = \Delta d_p/c = v d_p/c^2 = M_s t_p$, $t_p = d_p/c$, $M_s = v/c$. For example, one way propagation path between fixed source-observer of separation d=3m travelling on an aircraft at a speed of $M_s = 10^{-6}$ (670 miles/hour) and $c = 3x10^8$ m/s, gives $\Delta t_{pta} = M_s t_p = M_s d/c = 10^{-6}/10^8 = 10^{-14}s$. Using an interferometer to measure small time differences with a red laser $\lambda = 6x10^{-7}m$, $f = 5x10^{14}$ Hz. The number of instantaneous interference fringes (2π radians) $N = f\Delta t_{pta} = 5x10^{14}x10^{-14} = 5$, which is well detectable.

8 Initial claims

8.1 Basic physics

i) NR challenges SR as a more consistent and comprehensive theory. It is derived from the medium based ME's and the LT, it restores the continuity between classical and modern physics.

ii) All observed motional effects, both classical and relativistic, are with respect to the propagation medium. EM motional, gravitational and inertial fields propagate using the same medium.

iii) EM fields (light) are causal (predictable), they are solutions of the wave equation. Einstein's ether-less predictions are not a solution of the wave equation they are non causal (non predictable).

iv) Relativists who claim to have verified Einstein's ether-less relativity have usually verified Lorentz's medium based transform with its time and structural contraction through motion relative to the propagation medium.

v) Lorentz's rectangular transform axes predictions are causal; they predict motional properties. Oblique axes, which simulate Einstein's ether-less properties, do not represent a solution of the wave equation.

8.2 Moving systems and media

i) Einstein's Inertial Frame (EIF), which gives invariant ether-less properties, it is not a solution of the wave equation. All causal theories are variant; predicting Propagation Time Asymmetry (PTA) around moving systems.

ii) At Earth speeds, gravitational strengths and small integration times, relativistic and gravitational effects cannot contribute significantly to the dominant instantaneous classical PTA.

iii) Besides predicting the same measured observations as SR, NR also predicts additional measured observations that SR cannot predict, and establishes ether-less aspects of SR that cannot be measured.

iv) NR shows that not only does the medium exist over all space, but it is inhomogeneous. It is attracted to and compressed, both time and space, around gravitational bodies, according to Schwarzschild.

v) Einstein's homogeneous ether-less universe is not supported by measurement. NR supports a confirmed Gravitational Entrainment Model (GEM), describing media motions around the Earth.

8.3 Motional properties

i) NR allows distinction to be made between moving medium frames, and distinction between moving sources and observers with respect to the propagation medium, not possible without a medium.

ii) Source and observer motion, relative to the medium, contracts their time and structure. Observer contraction relatively expands the medium reducing gravity's time and space compression.

iii) NR shows time travel is non causal, it is not a solution of the wave equation. It is possible to travel to the past, causally, but not to interfere, it has already happened. It is not possible to travel to the future, by definition, it has not yet occurred. However, it is possible to slow ones time down through gravity and high speed motion.

iv) NR shows that it is possible to travel faster than the speed of light across hybrid frames, making intergalactic exploration possible. Distance is measured in the medium and time in the moving frame.

v) NR shows that LT is fundamental, SR is incomplete and its ether-less interpretation non causal. Attempts to explain observations without a medium lead to non causal predicaments.

8.4 Gravitational attraction

i) NR extends van der Waals near field electrostatic theory of attraction between molecules to account for the far field attraction between matter, constituting gravity in Einstein's GR.

ii) NR shows unsteady electrical fields are electromagnetic, steady dissimilar difference electrical fields are gravitational and residual difference electrical fields within the universe, are inertial.

iii) The medium has attractive 'dark' energy from attractive gravitational mass and possibly repulsive 'dark' energy through expansion of the medium, overcoming global gravitational attraction.
iv) All EM wave theories can be accounted for using a propagation medium. Einstein's predictions that are measurable are medium based, those that cannot be measured are usually ether-less.
v) NR provides a medium link between LT, SR, GR, PTA, inertial and accelerating frames. It also provides an optical description of the time Equivalence Principle, and a link to a unification theory.

The new propagation model around planets is described in Section 6, Chapter II. Einstein's ether-less non causal model is described in Section 3, Chapter III. The Lorentz transform is derived graphically in Section 4, Chapter III. The wave equation extension to the LT, for source and observer motion, is developed in Chapter V and summarised in Section 7, Chapter II. Applications illustrating the new medium based space-time are considered in Section 2 and Section 4, Chapter VI. The nature of gravity is considered in Section 2, and propagation around planets considered in Section 3, Chapter VIII.

9 Conclusions

1. The Lorentz Transform (LT) is the fundamental EM motional theory. It is based on a propagation medium (ether). It is causal, it predicts EM motional effects.

2. All causal (predictable) wave theories require a medium, including EM theories. Maxwell's equations, LT and the EM wave equation are all medium based.

3. All motional effects are with respect to the medium:
   (a) Fundamental classical PTA effect (propagation time asymmetry).

(b) High speed relativistic effect (Lorentz's time and structure contraction).

4. Einstein's measured SR predictions are medium based:
    (a) Can be derived directly from classical medium based wave theory.
    (b) Can show that Einstein's field equations (wave equation) are medium based.

5. Einstein's ether-less SR is non causal. It is based on Einstein's invariant inertial frame, where the speed of light and mechanics are invariant. However, it has no medium to create variant PTA or communicate with the observer.

6. Einstein's medium based SR is causal but incomplete, it does not distinguish between:
    (a) Measured source and observer motional differences.
    (b) Stationary and moving medium frames.

7. New Relativity (NR) extends LT to distinguish between source and observer motion and between stationary and moving medium frames, reuniting classical and EM wave theories.

8. Relativists attribute EM wave propagation to space-time, without definition. Space in NR is filled with a propagation medium, defining position. Time is a measure of the rate of happening, order and spacing of the observed events.

9. A difference electric field from a finite distribution of dissimilar charges (dipoles) within atoms and molecules creates an attractive field constituting gravity.

10. The propagation medium's existence provides a common basis for relating LT, gravity and accelerating frames.

Chapter II:

Characteristics of New Relativity

1 Introduction

Accepting the presence of the propagation medium (ether) for the propagation of electromagnetic (EM) waves (light) explains the universe in a rational manner. To investigate what constitutes an EM source and its propagation medium, in an elementary way, consider an assembly of electrons on a metal sphere. The resulting charge produces an electrostatic field, which is capable of action at a distance, as demonstrated by the diverging leaves of a charged electrometer. Or the frightening sight of hair standing on end close to a charged Van de Graaf generator, as illustrated in Figure 2.1(a). Or simply combing one's hair on a dry day.

The electric field requires a propagation medium to propagate its influence to distant places. If the charge is varied the electric field will vary, transmitting EM disturbances along its electric field lines. The disturbances travel at the speed of light, as predicted by Maxwell's (1865) medium based electromagnetic field equations (ME's). If the charge is modulated at a high enough frequency, then the electromagnetic disturbances will propagate as radio waves that can be heard on a radio, as shown in Figure 2.1(b). At still higher frequencies, the waves become visible as light, considered as waves or a burst of discrete photons travelling at light speed, as depicted in Figure 2.2(a).

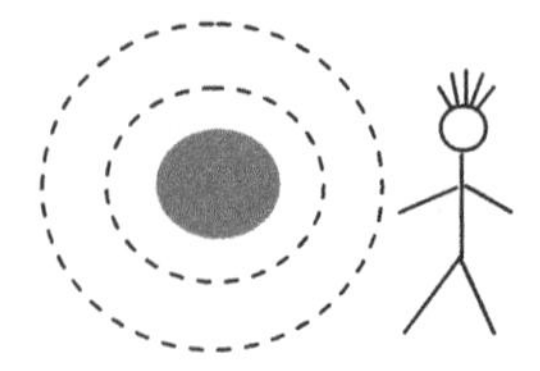

(a) Steady charge generates an electric field that can produce action at a distance (make hair stand on end)

(b) Varying charge generates modulated electric field (EM waves) that can be heard on radio

Figure 2.1 Examples of electric and electromagnetic fields propagating in the propagation medium

If a constant charge or an electron is now moved relative to the propagation medium, or an observer moves relative to a fixed charge and its field, a field strength change (disturbance) will be detected. If our steady charge or electron is moved close to the speed of light in a vacuum, and then released in air, an impulsive wave is formed around the moving charge in the more dense, lower speed medium. Cerenkov's (1934) impulsive radiation can actually be seen as a blue light in a nuclear reactor, or as a high speed electron exits an accelerator.

Here a 'super-lightic' bow wave is formed in a similar way to the bow wave from a ship exceeding its water wave speed, or a supersonic boom in the case of a fast moving aircraft, as illustrated in Figure 2.2(b). Further, discrete light (photons) is emitted when electrons combine with anti-electrons, or electrons jump orbit rotating around an atomic nucleus, or through simple electrostatic discharge as in lightning etc. Thus photons of light and radio waves can both be considered as EM disturbances using the basic electric field as a conveyer of EM waves. If there was no propagation medium, there could be no propagating electric field and no transmitted EM disturbances. None of the effects illustrated in Figures 2.1 and 2.2 could happen without a medium, Einstein believed they could.

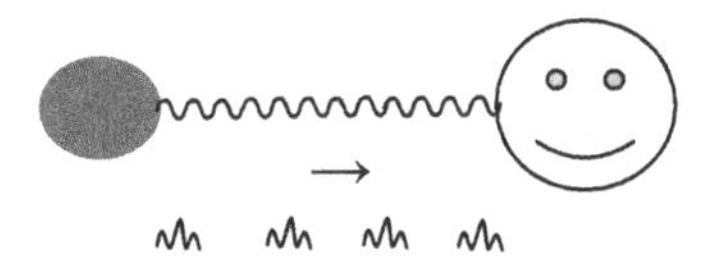

(a) At very high frequencies an EM disturbance can be considered as a continuous light wave or a burst of discrete photons

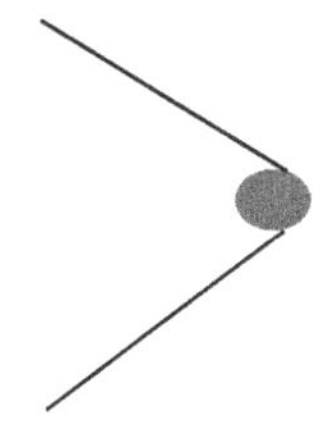

(b) A high speed electron entering a more dense medium can generate a 'super-lightic' bow wave, similar to a super-sonic boom

Figure 2.2 Further examples of electric and electromagnetic fields propagating in the propagation medium

2 Relative Motion

Consider now the case of inertial frames (systems moving at constant velocity). Galileo (1632), was the first official documented comment concerning the lack of motional effect in a moving galleon, on a calm sea. Notionally, one can play ping-pong, pour a glass of wine or perform an experiment on a smoothly running underground train without noticing its motion, as depicted in Figure 2.3. In the case of two trains in a tunnel; if one starts to move, it is not obvious to opposite passengers on each side of the two trains, which train is moving, only relative motion seems to be important. For car users, the same effect can be experienced in an automatic car wash. The effect is quite impressive, it is not obvious which system is moving.

According to Einstein's invariant, inertial frame, constant motion was not considered to be detectable, for example viewing across the two trains moving together at the same speed. They could be moving or stationary, apart from slight movement from the uneven tracks. Mechanically, there is no difference between a stationary and a constantly moving system, as Galileo observed. Einstein was so impressed by the relativity effect that he developed his theory of relativity. His belief was based directly on relative motion between systems, rather than motion relative to the propagation medium that transmitted the observations. Einstein thought the medium was redundant, and that observations could propagate without a medium.

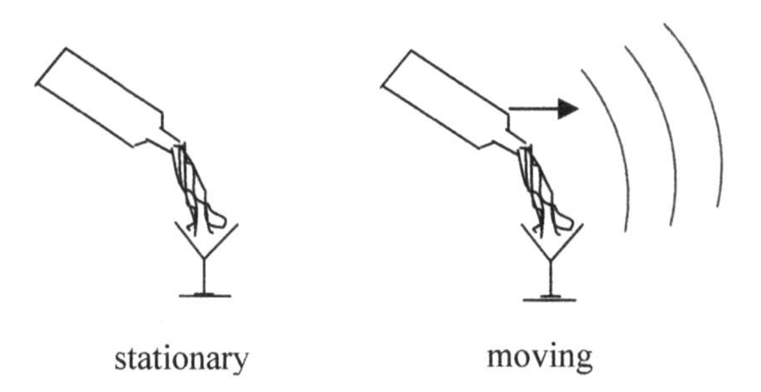

Figure 2.3 Mechanically there is no difference between a stationary and a smoothly moving system. However, moving in a medium, there is a difference optically, the waves compact in front of the source, progressively increasing with speed

However, there is no experimental data confirming Einstein's assumptions. To keep the universe tidy, causal and consistent with classical theory, history tells us that to observe these motions, they should be relative to something more definite - their propagation medium, according to the medium based theory of Lorentz, described in detail in Section 4, equation 3.11, Chapter III. The system that moves with respect to the medium will then show motional effects, such as Propagation Time Asymmetry (PTA), demonstrated in Figure 1.1 and in front of the source indicated on the right of Figure 2.3. At sufficiently high speed, these individual motions will become obvious, optically, as in visible Cerenkov radiation, illustrated in Figure 2.2 (b). Without a medium there could be no such motional effects, the light paths would be indeterminate, there would be no light.

It was therefore expected that the Earth's motion through the heavens should show some small optical effect. However, in the Michelson and Morley Experiment (MMX) (1887), shown in Figure 1.2, the light round trip propagation time (RTPT), reflected by mirrors in the direction and perpendicular to the direction of the Earth's motion, was found to be Propagation Time Symmetrical (PTS). This supported the null effect of Galileo's observation. It was a major set-back for the ether supporters. It was expected that there would be PTA, as illustrated in Figure 1.1 and predicted by equation 1.2, for an object moving relatively to its propagation medium. The result appeared to be against all intuition and scientific principles, and yet there is a perfectly logical explanation.

It is shown that the PTS in the MMX is caused through the medium moving with the Earth, as proposed by Stokes (1845), explaining the null result in a natural way. Had Einstein measured the one way propagation time (OWPT) in a frame moving relative to the Earth and its 'stationary' medium, as Sagnac (1913) did, depicted in Figure 1.3, and discussed in detail in Section 5.1, Figure 4.4, Chapter IV, Einstein would have measured PTA with respect to the propagation medium. Einstein's ether-less relativity would then have had to be abandoned. However, using the RTPT method, there was not enough sensitivity to distinguish between PTA and PTS, if there had been a difference. The OWPT method is considerably more sensitive than the RTPT method, by over a million times, as discussed in Sections 6.1 and 6.2, Chapter I. It is well capable of detecting PTA and therefore constant motion relative to the medium.

It appears that the MMX symmetry had a big influence on Einstein in abandoning the propagation medium, although Einstein claimed not to have heard of the MMX in 1905, performed many years earlier. He proposed to his Professor at the Zurich Polytechnic (Weber 1899), a similar experiment to the MMX, but was told that many others (13) had already documented a negative result, including the MMX. Einstein probably learned of the MMX through Lorentz's (1899) book 'Attempt at a theory of electrical and optical phenomena in moving bodies', well known to scientists at that time, describing the MMX.

3 Residual Gravitational Field

Turning now to accelerative motion, it is obvious and pretty special. Stopping, starting, swerving, rotating, bobbing up and down are all obvious changes in speed and direction. Newton (1687) declared that

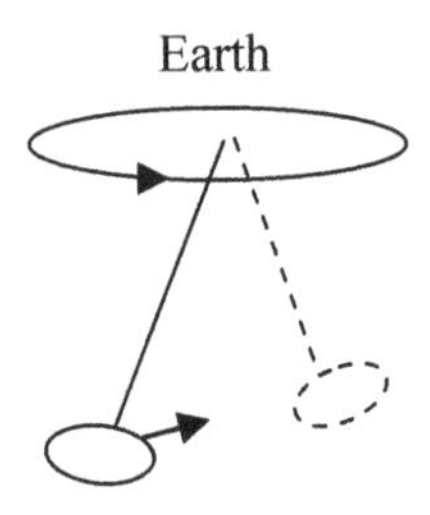

Figure 2.4 Foucault's swinging pendulum plane is not affected by rotation of the Earth

acceleration was with respect to absolute space. Foucault (1850), nearly 200 years later, through his long, freely suspended pendulum, illustrated in Figure 2.4, demonstrated that its oscillating plane was independent of the Earth's rotation and probably its orbital motion. It appears that the Earth's gravity is sufficient to drag the gravitational atmosphere with its rotation, close to its rotating surface, and attract the mass-less medium generally around the planet to many Earth diameters, as discussed in Section 6, Figure 2.8. However, the Earth's gravity and rotation do not appear to affect the heavy pendulum's plane of oscillation, which seems to be influenced by an inertial field, as discussed in Section 2.7, Chapter VIII, transmitted by the medium at rest in space.

Mach (1887), probably encouraged by Foucault's pendulum demonstration, came to the conclusion that acceleration was relative to the total mass of the distant universe. Thus, those of us who are natural accelerometers when travelling, becoming nauseous through accelerative motion, should feel special. It appears that we are reacting to the total mass of the universe, through its residual gravitational field. Mach's conclusion provokes the almost unbelievable thought experiment. It's possible to support a non spinning top on its tip by rotating the universe, as illustrated in Figure 2.5.

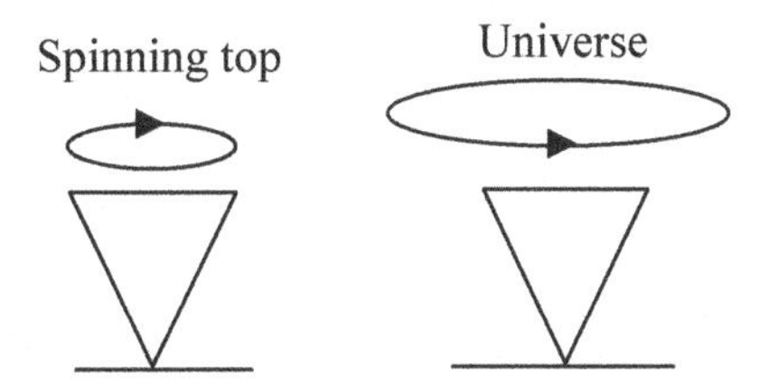

Figure 2.5 To keep the top upright it does not seem to matter which rotates, the top or the universe, providing there is relative motion between the top and the inertial gravity field

If there was no relative motion, or indeed no inertial field, then the top would 'fall over'. Thus there appears to be an overriding universal inertial gravitational field controlling the pendulum's plane of oscillation and supporting the spinning top. The gravitational field from all directions in space, from all the gravitational matter in the universe, appears to provide such a finite inertial field with no net force but with finite energy. This assumes the mass contributing to the field, at any point in space, increases according to an expanding surface area of a sphere (square law). In this way the inverse square law field decay from the gravitational mass will be compensated. If this field is basically uniform over all space, then the inertia of the top would be unchanged anywhere in the universe, on a planet or outer space. Whereas, a pendulum's period is dependant on local gravity.

4 Classical and EM Systems

Against popular belief, it is shown that all moving classical sources (acoustic, vibration, water waves, etc.) do have similar properties to electromagnetic (EM) light systems in motion. The only difference is that in optical systems Lorentz's time and structure contract through high speed motion through the medium, where the medium is generally at rest in space, or moving with gravitational bodies. Causally, this results in the speed of light being invariant, but nothing more sinister. Both require a propagation medium, exhibit Propagation Time Asymmetry (PTA) and are both causal. It is helpful to consider ordinary everyday classical systems, whose high speed effects are readily observed.

For example, a moving acoustic observer with respect to a stationary propagating medium (air), has a slightly different motional Doppler (1842) effect (observed frequency change) than a moving acoustic source as illustrated in Figure 2.6 (a). However, observations of an acoustic source and observer, of fixed separation, in a frame moving with the medium (air), as in a conversation in a moving closed system, obviously provides no change in the acoustic properties (zero Doppler effect). Here, the medium moves with the system giving the same forward and reverse propagation times (simultaneity), irrespective of the system speed. This is a similar effect to what Einstein believed happened within his inertial frame, except he believed that there was no propagation medium.

Also, a fixed acoustic source and observer moving at constant speed relative to the propagating medium, as in a moving open system, again results in zero Doppler effect. Here the source and observer Doppler effects cancel giving zero frequency shift with change in system velocity. *However, the forward and reverse acoustic propagation times in the medium do change, they become unequal (asymmetrical) and vary with system speed,* clearly revealing motion relative to the propagation medium. The asymmetry is illustrated in Figure 2.6 (b) and (c), physically realised in Figure 1.1, and predicted using equation 1.2, in Chapter I. The asymmetry (PTA) provides an absolute method of measuring the systems speed in the moving frame, without external measurements (looking outside the system).

(a) Classical sources and observers in motion have different Doppler effects for the same speed, whereas EM systems are identical

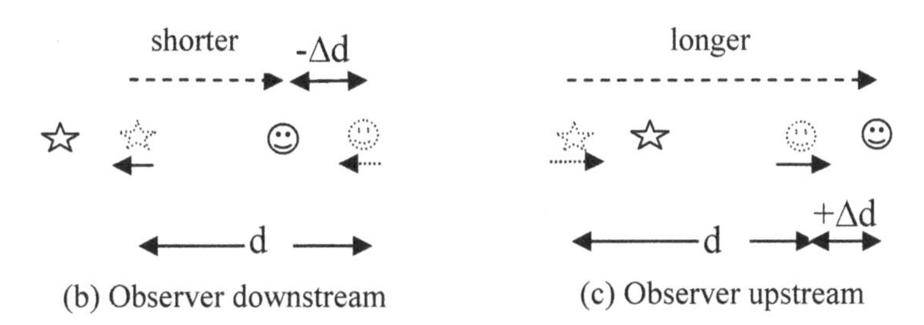

(b) and (c) Fixed source and observer moving relative to the medium gives Propagation Time Asymmetry (PTA) revealing their motion. Observer upstream has the longest propagation distance

Figure 2.6 Unlike classical systems, EM sources and observers in motion have identical Doppler effects. However, both classical and EM systems have identical PTA, illustrated in Figure 1.1. EM systems have additional Lorentz contraction of space and time at high speed

For EM systems in motion, the Lorentz Transform (LT) describes the complete motional effect. Both the classical Propagation Time Asymmetry (PTA) above, and the additional Lorentz Contraction (LC) created at high speed, are predicted, incrementally by equation 1.1, Section 1, Chapter I, i.e. $\Delta LT = \Delta PTA + \Delta LC$. The PTA is present in all wave theories, including Lorentzian. For EM individual source and observer motion, the Lorentz contraction factor α in the LC modifies the Doppler factor ε in the PTA, in equation 2.2, in Section 7. This now makes the EM source and observer Doppler effects identical, as in equation 2.4, unlike classical theory given by equation 2.5.

Also, EM Doppler effects are unchanged between a fixed source and observer through motion, equation 2.3. What is more important is that the PTA in and against the direction of motion, shown in Figure 2.6 (b) and (c), remains asymmetrical, given by equation 1.2, similar to classical systems. *So, according to Lorentz, the PTA is a fundamental characteristic of EM systems in motion relative to the medium.* Whereas, Einstein's ether-less inertial frame simulating Propagation Time Symmetry (PTS), analysed in Section 3, Chapter III, is inappropriate. It is unable to resolve the situation, it cannot detect constant motion, nor predict PTA, which is observed in the real world.

5 Moving Carriage

For a more interesting example consider the effect of reflected light, as discussed in detail in Section 4.2, Figure 7.4, Chapter VII. Here, EM propagation in the direction of motion, as in the light reflections from a mirror on a carriage wall, perpendicular to the train's motion, is considered. Again the EM medium is at rest on the Earth's surface. The EM wave propagation time in and against the direction of motion will be asymmetrical and variant. Similar to the classical case, establishing absolute motion, in contrast to Einstein's invariant inertial frame.

However, the reflection propagation time from the window aligned with the train's direction, as illustrated in Section 4.1, Figure 7.3 Chapter VII, becomes independent of motion, unlike the classical case. Amazingly, through the LT, the propagation time becomes invariant perpendicular to the direction of motion. Here the increase in propagation time (geometric diagonal path increase by α^{-1} compared with the perpendicular path for a stationary system) is exactly offset by the relativistic observer time slowing α.

Thus in the EM case, one's reflected image propagation delay from the mirror on the perpendicular wall will be dependent of the train speed. But looking at the reflected image from the window, the delay will be independent of the train's motion. This is a second subtle difference caused by the LT between EM and classical systems. The first involves equal Doppler effects for individual EM source and observer motion. Apart from minor fascinating differences, it can be seen that classical and EM systems motional effects are basically similar, at low speeds they become virtually identical.

Further, consider a bright light flashed at the centre of our stationary carriage, inside and on the roof, so that it would be seen simultaneously at each end of the carriage. And also, seen in each direction of the stationary rail track, for the same distance apart, as illustrated in Figure 2.7(a). For a moving carriage, the light would of course still be seen simultaneously on each side of the stationary track. *But, according to Lorentz, Figure 2.6, equation 1.2, and in detail in Section 3, Figure 3.2, equation 3.11, Chapter III, there will be propagation time asymmetry (PTA) at each end of the carriage moving relative to the Earth and its medium, as indicated in Figure 2.7(b).*

PTA is fundamental, it is the time domain asymmetry equivalent to the Doppler asymmetry in the frequency domain. It was established a long time ago by Sagnac (1913), discussed in Section 2, Figure 1.3, Chapter I and described in detail in Section 5.1, Figure 4.4, Chapter

IV. This asymmetry and variance is, of course, in complete disagreement with Einstein's ether-less inertial frame, which without a medium cannot distinguish between a stationary and constantly moving system. Denying the medium's existence, effectively removes the classical PTA part of the LT, artificially creating a discontinuity between classical physics and modern motional optics.

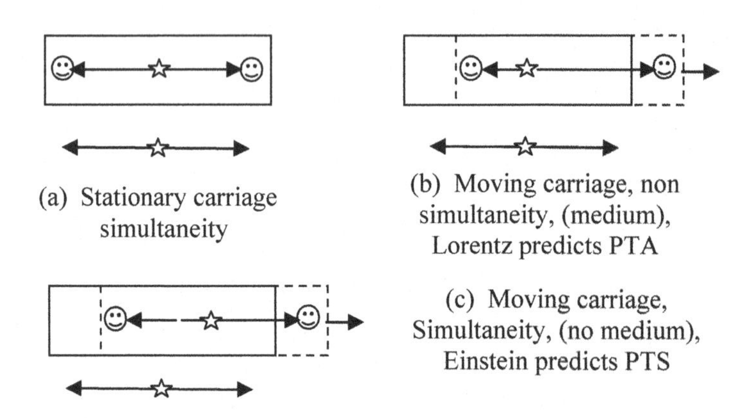

(a) Stationary carriage simultaneity

(b) Moving carriage, non simultaneity, (medium), Lorentz predicts PTA

(c) Moving carriage, Simultaneity, (no medium), Einstein predicts PTS

Figure 2.7 Light flash at centre of stationary and moving carriage with and without a propagation medium

Einstein's ether-less non causal inertial frame (EIF), considered in Section 3, Chapter III, is illustrated in Figures 2.7(c). With the assumption of no medium, and its invariant propagation time symmetry (PTS), there is no difference between a stationary and moving frame. *This ether-less PTS concept has never been measured or verified. It is in direct opposition to the measured asymmetry and varying LT prediction, and is therefore not supported by Lorentz's medium based theory.* Amazingly, this ether-less explanation, not supported by basic physics, was accepted and continued to be accepted, mainly because of its apparent support from the Michelson and Morley Experiment (MMX) (1887).

In the MMX, PTS upstream and down would have become asymmetrical if the Earth had moved relative to its surrounding medium.

However, the propagation time symmetry (simultaneity) in the stationary medium moving with the Earth's surface was interpreted as supporting a belief in Einstein's concept of relativity, instead of the medium moving with the Earth, according to Causal Condition I, Section 1.4, Chapter I. This violation against basic wave theory (relative motion without PTA), created considerable mysticism surrounding SR. It undermined confidence in what is a straight forward understanding of wave propagation around systems in motion.

Whether or not the signals arrived simultaneously at each end of the moving carriage, Einstein never knew. This important experiment could not be performed through lack of sensitivity, even if there had been relative motion with respect to the medium. Using the round trip propagation time (RTPT) of the MMX, the propagation time difference in each direction almost cancels, severely reducing the sensitivity. However, this optical method is now easily do-able, using a one way propagation time (OWPT) measurement, described in Section 6.1, Chapter I. For typical speeds on Earth, the OWPT method results in a sensitivity increase over the RTPT of the order of 10^6, enabling this one way method to detect slow motion on Earth relative to the 'stationary' propagation medium, such as the motion of a train.

.

6 Gravitational Entrainment Model (GEM)

An abundance of available data, Wright (2010)[1]- (2013)[10] and supported by references listed in this book, provide adequate proof of the medium's presence. The model that appears to fit all known data is the medium based Gravitational Entrainment Model (GEM), illustrated in Figure 2.8. This establishes that moving media should replace Einstein uniformly empty space.

Bradley's (1725) stellar aberration angle described in Section 4.3, Figure 7.5, Chapter VII is caused through the Earth and telescope (**observer**) orbiting around the Sun, moving laterally to the light from a distant star propagating in the medium at 'rest' in space. Its maximum **forward** angle from the star direction is given by $\varphi=\tan^{-1}(M_{orb})$, where v_{orb}=30km/s is the Earth's orbital speed, $M_{orb}=v_{orb}/c=3x10^4x 3^{-1}x10^{-8}=10^{-4}$ is the orbital Mach number, which gives a very small, but measurable angle. The medium with no mass and unity refractive index surrounding the Earth, does not affect the aberration angle. The angle is formed in the medium, rather than being an apparent resolved angle, without a medium.

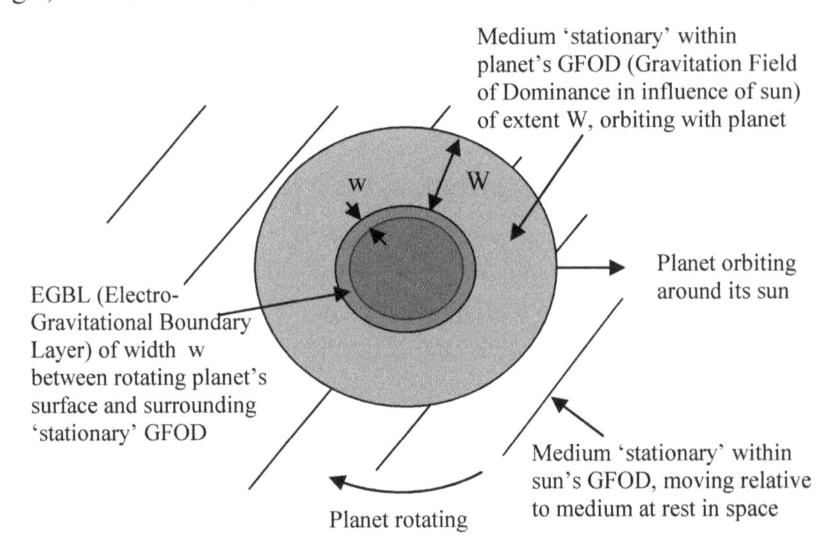

Figure 2.8 Gravitational Entrainment Model (GEM). Medium profile around a rotating and orbiting planet, according to Michelson and Morley, Sagnac, Michelson and Gale, Penzias and Wilson, Hafele and Keating, Saburi et al, Reasenberg et al, COBE and GPS.

Whereas, Global Positioning Systems, (GPS) (1992), discussed in Section 3.5, Figure 8.5, Chapter VIII, gives a **backward** displacement angle, normal to the direction of motion, formed by satellite (**source**) motion relative to the 'stationary' medium surrounding the Earth. This angle is given by $\varphi=\tan^{-1}(M_{sat})$, where M_{sat} is the satellite Mach number. At the equator the rotational surface speed is $v_{sur}\approx$440m/s, $M_{sur}=v/c\approx(440/300)x10^{-6}\approx1.5x10^{-6}$. These examples clearly distinguish between source and observer motion relative to the medium, which Einstein's ether-less SR cannot distinguish between. For a typical GPS satellite at $d \approx 20.10^6$m above the Earth's surface, transmitting signals to the Earth, and rotating relative to the stationary medium surrounding the Earth, creates a one way PTA displacement on Earth.

From equation 1.2, $\Delta d_{pta}=vt=Md$, where t=d/c is the propagation time between the satellite and Earth station. Taking the Earth's surface rotational speed $M_{sur} \approx 1.5x10^{-6}$, gives a measured surface displacement of $\Delta d_{pta}=1.5x10^{-6}x20x10^{6}=30$m. Further confirmation is through Saburi et al (1976) who investigated satellite communication across the Pacific, discussed in Section 3.7, Figure 8.6, Chapter VIII. They measured a predicted PTA=2Mt=300ns (100m), in and against the direction of the geostationary satellite rotating with the Earth, relative to the 'stationary' medium surrounding the Earth. The factor of 2 is for propagation to and from the satellite to the ground station.

Also Hafele and Keating (H&K) (1972), discussed in Section 5.7, Figure 4.6, Chapter IV, used atomic clocks to measure time change through motion. They flew for 24 hours in a aircraft at 10km above the equator. Asymmetric propagation time slowing of -273ns/day and 59ns/day relative to the Earth's surface time, in and against the direction of the Earth's rotation were found. This asymmetry corresponds to motion relative to the stationary medium surrounding the Earth, not relative motion between the aircraft and the rotating Earth's surface, as discussed in detail in Section 5.7, Chapter IV.

The orbiting medium with the Earth is shown to move relative to the 'stationary' medium surrounding the Sun and Solar System, moving through the universe. This has been confirmed through symmetrical signals, in each direction (effectively no PTA), between Mars and Earth in the Mars-Viking Lander project by Reasenberg et al (1979). If the medium had not moved with the Solar System, but was stationary in space, there would have been a considerable PTA in communication time, in and against the direction of the Sun and Solar System motion ($M_{sun}=v_{sun}/c=10^{-2}$) through the universe, which is not measured.

The predicted 30m and 100m measured in the PTA examples above, are calculated using only classical medium based theory. Relativistic motional effects are negligible in these cases unless integrated over long periods of time. *If the surrounding medium had rotated with the Earth, down to its surface, neither of the signal PTA's would have occurred.* These classical predictions indicate that the medium is attracted to and surrounds large gravitational bodies to an extent ‘W’. This model, with the surrounding medium 'stationary' with the orbiting Earth, is supported by the latest results from NASA’s Gravity Probe B (2011).The extent appears to be controlled by the planet’s Gravitational Field of Dominance (GFOD) in the presence of the Sun’s gravitational field of influence, discussed in Section 3, Figure 8.2, Chapter VIII. In the Earth’s case, the effective GFOD boundary appears to be approx-imately W=50 Earth radii.

The MMX (1887), described in Section 2, Figure 1.2, Chapter I, demonstrated that light propagation on the Earth’s surface, through a fixed optical system, was unaffected by its motion through the universe (medium moved with the Earth). Sagnac (1913), described in Section 5.1, Chapter IV, showed that motion of his rotating mirrors, relative to the Earth’s surface (and medium), in and against the light direction, caused PTA exactly according to equation 1.2. Michelson and Gale (1925) described in Section 5.3, Chapter IV, determined that the medium clings to the Earth surface moving progressively faster from the poles to the equator. This causes a measured difference in light propagation time (speed) over the same distance, at different latitudes on the Earth’s surface, revealing the Earth’s rotational speed relative to the surrounding ‘stationary’ medium.

These last three descriptions establish that the medium on the Earth’s surface exists and rotates with the Earth relative to the ‘stationary’ medium surrounding the Earth, within its GFOD. Again these effects are based on classical physics, the relativistic effect is negligible, unless integrated over a considerable period of time. This data is explained through an Electro-gravitational Boundary Layer (EGBL) of extent ‘w’ between the rotating surface and the ‘stationary’ medium surrounding the Earth. The extent of the boundary layer on the Earth is less then 10km, according to Hafele and Keating (1972). Transmission path variations can occur passing through the ionosphere positioned between 100km-1000km above the Earth.

In the absence of gravitational matter, the medium in the universe appears to be stationary, on average, providing a universal reference for motion. The cosmic microwave background (CMB), detected by Penzias and Wilson (1965), is shown to be EM radiation, propagating uniformly in all directions, throughout the universe, relative to the propagation medium basically at rest in space. The stationary medium has also been confirmed through the Cosmic Background Explorer COBE (1992). Here the CMB energy collection increases with system motion relative to the stationary medium, similar to trawling fish nets in any direction catch more fish than stationary ones.

A typical SR and GR analysed data set, analysed by Kramer et al (2006), can be described by the medium based NR. Kramer et al studied pulses from a distant double pulsar system. The pulsar’s orbital time slowing and orientation change is attributed to Lorentz's time contraction of the high speed pulsar relative to the medium. The propagation time delay is caused through the intense gravity of the heavy pulsars (neutron stars) compressing the medium according to Schwarzschild (1916), retarding the propagation in the vicinity of the pulsars. None of this activity can be attributed to ether-less relativity between systems, as Einstein believed.

7 EM Motional Analysis (EMMA)

The complete solution of the motional wave equation for arbitrary sources and observers in motion is described in detail in Chapter V. In order to give an overview and a working knowledge of the new theory, distinction is made between source and observer time relative to the medium time. Three time and space scales are essential and specified (source, observer and medium). Whereas, only two un-specified scales are used in Einstein's incomplete SR, where there is no distinction made between source and observer motion. The segment of integrated or accumulative observed event time τ_o, in terms of the integrated source event time τ_s, for arbitrary source and observer flight paths through the universe, is from Section 4 equation 5.28 and equation 5.31 in Chapter V, given by:

$$\tau_o = \int [\varepsilon_s \alpha_s^{-1} \varepsilon_o^{-1} \alpha_o]_{tp}\, d\tau_s, \qquad t_o = \tau_o + R_o/c, \qquad R_o = \alpha_o R_p \qquad (2.1)$$

The situation is illustrated in Figure 2.9 for arbitrary motion. The observed events depends on the initial propagation distance R_p in the propagation medium (R_o in the moving frame). Also on how the individual motional operators ε_o^{-1}, α_o, ε_s, and α_s^{-1} change with time during the source and observer journeys through space. In this medium based theory, time changes instantly at the source emission and the later observer reception times, according to their speeds in the medium at that instant, not according to relative motion between systems. At any source emission/observer reception instant, the corresponding instantaneous time transform is given by:

$$K = \tau_o/\tau_s, = \varepsilon_s \alpha_s^{-1} \varepsilon_o^{-1} \alpha_o \qquad (2.2)$$

The Lorentzian or relativistic omni-directional time change operators $\alpha_s^{-1}\alpha_o$, given by $\alpha_s=(1-M_s^2)^{1/2}$ and $\alpha_o=(1-M_o^2)^{1/2}$, result in instantaneous Lorentzian time slowing at a moving source and effective time quickening at a moving observer. Their effects are permanent. The classical Doppler factors $\varepsilon_s \varepsilon_o^{-1}$, given by ε_s=1-M_s cos σ_s and ε_o=1-M_o cos σ_o, compress or expand time locally, depending on whether the systems are approaching (Mcos σ positive for source and negative for observer) and vice versa for receding systems. The motional angles σ are formed with a straight line between the source emission point and observer reception position, illustrated in Figure 5.1, Chapter V. These effects are temporary, when the motion stops ε_s and ε_o become unity. In a closed loop (flight path) the total contraction and expansion are equal (they cancel).

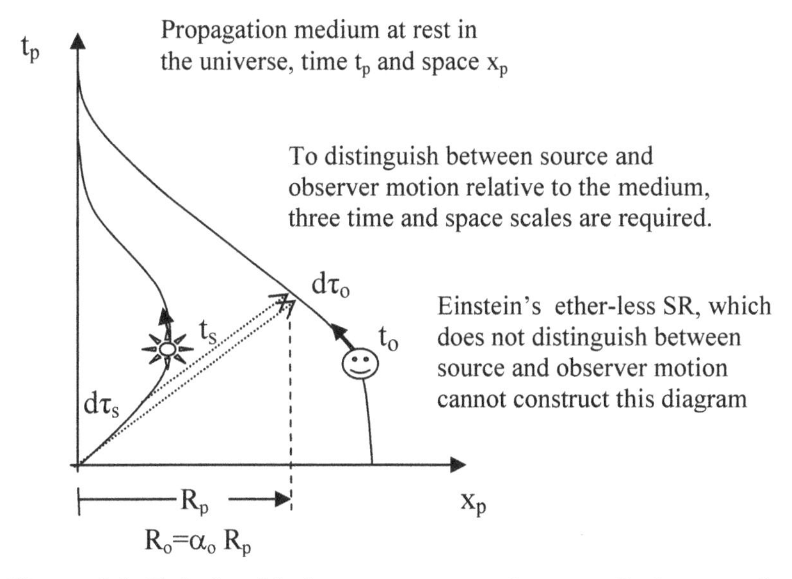

Figure 2.9 Relationship between source time t_s and observer time t_o for arbitrary motion relative to a stationary medium of time t_p

The vertical and horizontal axes in Figure 2.9 represent absolute time t_p and space x_p, in the medium, respectively. Corresponding times, between the source emission event time τ_s and observer reception event time τ_o curves are connected by propagation light paths. These paths represent light moving forward with time, from source to observer, going from left to right in the direction of the arrows, whose gradient is 1/c (45° if x_p is in light years). In these particular flight path examples, the τ_s and τ_o time curves start at the same time but at a different place. They end at a different time, but at the same place. Interchanging the source and observer for the same flight paths will give different observations. The light paths then go from right to left and have a negative gradient, giving a different observed time history for the same flight paths, again not predicted by Einstein's SR.

For source and observe motion in the same straight line of fixed separation and equal velocities s=o, $\varepsilon_s = \varepsilon_o$, $\alpha_s = \alpha_o$ from equation 2.2:

$$K_t = \tau_o/\tau_s, = K_s K_o = \varepsilon_s \alpha_s^{-1} \varepsilon_o^{-1} \alpha_o = 1 \qquad (2.3)$$

i.e. the source and observer Lorentz contraction and Doppler effects cancel showing no sign of observed motion in the moving frame.

However, in the case of an individual approaching EM source and observer $M_s = -M_o$, and $\alpha_s = \alpha_o$., then from equation 2.2:

$$K_s = \varepsilon_s \alpha_s^{-1},\ K_o = \varepsilon_o^{-1} \alpha_o,\ \ K_s/K_o = \varepsilon_s \varepsilon_o / \alpha_s \alpha_o = 1$$

as (2.4)

$$(1-M)(1+M)/[(1-M^2)^{1/2}]^2 = (1-M^2)/(1-M^2) = 1$$

i.e. approaching EM source and observer observations are identical. Whereas for the classical case $M_s = -M_o$, $\alpha_s = \alpha_o = 1$

$$K_s = \varepsilon_s,\ \ K_o = \varepsilon_o^{-1},\ K_s/K_o = \varepsilon_s \varepsilon_o = (1-M)(1+M) = (1-M^2) \qquad (2.5)$$

i.e. K_s and K_o are not identical. Thus the observed effect of an individual approaching EM source and observer is identical, unlike the classical case. The Doppler effects are modified by the Lorentzian contraction factors producing the exact equality. Applications of the new motional optics are considered in Figures 6.1, 6.2, 6.6 and 6.7.

In the new theory, time can be quickened or slowed, according to the speed and direction of the source and observer relative to the medium.

Source motion slows time, and observer motion slows observer time, relatively increasing the observed time of its surroundings. This allows observer motion to reduce the time slowing process of sources in motion or through gravity. Figure 2.9 cannot be constructed, according to Einstein's ether-less SR. It is valid generally for arbitrary (constant and accelerative) motion. Whereas, Einstein's inertial frame cannot distinguish between source and observer motion, or between constantly moving and stationary systems. With the added flexibility of three time and space scales, all known observations now appear to be accountable.

Also time travel (material transmission) into the future or past is not possible, it is not in agreement with causality. To actually participate, change the source events as they happen, the observer must come in physical contact (interact) with the source, *at the source event time.* By definition this rules out any kind of distant participatory time travel. Causally, it is possible to observe past events, but not to participate in them, because they have already happened. It is not possible (non causal) to observe or participate in source events in the future, because they have not yet occurred. Through motion or visiting a large mass, a traveller can slow his/her time down compared with a stationary system time and return to its future, actually participating in its present. This is quite feasible, but it is not reversible time travel. The process is one-way, one cannot reverse the flow of time in this situation. All that can be achieved is to alter the rate of ageing (time slowing) of one system compared to another.

8 Conclusions

1. To understand the mechanics of the universe, it is important to understand motion. It is self evident that to comprehend motion requires observations, and observations require light. Interpreting motion through light requires a knowledge of electromagnetic (EM)

waves (photons) and the existence of their propagation medium. These observations are predicted through solving the wave equation based on the propagation medium. Its solution is causal, it is valid for all kinds of waves, classical and electromagnetic. In the case of EM waves, Maxwell's medium based wave equation predicts the source events propagating through the EM medium, eventually seen in terms of light at the observer. *The reality of observed information through wave propagation, is that events cannot be observed, judged or predicted without a propagation medium.*

2. For stationary EM sources and observers there is no confusion or argument, a medium is naturally accepted, the same as classical systems. Without proof or robust scientific reasoning, Einstein decided that for motion, EM waves suddenly became different, and that a propagation medium was no longer required. This is not possible, it is against fundamental physics (motional optics), a medium is required to transmit waves and predict causal observations. According to basic wave theory, it does not matter whether the propagation medium or source and observer move. *It is the motion between the source and observer, relative to the propagation medium that affects the observations, not motion relative to each other, as Einstein believed.*

3. Maxwell's Equations (ME's) require a propagation medium for the solution of the EM Wave Equation (WE). The Lorentz Transform (LT) requires a propagation medium for the motional solution of its WE. LT predicts that system time and space contract through motion relative to the medium, by the same fraction. This makes the speed of light invariant, but not its Propagation Time Asymmetry (PTA), which remains variant in a frame moving with respect to the medium. *The medium based LT is fundamental in the solution of the EM motional wave equation and for combining velocities in the Relativistic Addition of Velocities (RAV).* LT is causal, the cause occurring before the effect, predicting motional observations through motion relative to the medium. ME's, LT, WE, and RAV are all medium based.

4. Without a medium there is no other known propagation mechanism. Einstein's ether-less relativity assumes no propagation medium, it therefore cannot be a solution of the wave equation. It is non causal, the effect can occur before the cause, it is irrational and cannot be measured. Although Einstein denied the existence of the propagation medium, it can readily be verified from inspection of Einstein's motional electrodynamics that he in fact used a propagation medium to solve his wave equation. *Einstein's field equations predicting measured SR, are based on Lorentz's medium based motional transform. Thus, Einstein's measured predictions are medium based, not ether-less, as he claimed.*

5. Without the propagation medium there is no universal reference, how can actual speeds be measured? The absolute speed of light would be in doubt, how could its speed be judged without its medium? How could systems in motion know they are moving, without the medium? How could their time and space scales know when to contract with motion according to Lorentz without a medium? How could the medium be compressed by gravity, both time and space, according to Schwarzschild metric. *Without a medium there would be no structural-time contraction of moving systems or propagation medium space and time compression through gravity.*

6. Contrary to Einstein's belief, in his Special Relativity, there is nothing mysterious or special regarding light propagation and EM systems in motion. Causal observations between source and observer motion can be predicted by simply solving the regular classical wave equation based on a propagation medium. *The only additional feature for optical systems, compared to the classical theory, is that time and space contract through motion relative to the medium.* Einstein's SR is therefore inconsistent, claiming no medium but then used one in his field equations. Reinstating the medium restores the link between classical and modern physics.

7. Further SR is incomplete, not recognising the medium's presence prevents distinguishing between source and observer motion, clearly evident in the observations. To complete the propagation process and to distinguish between source and observer motion, absent in SR, it is essential to extend Lorentz's motional wave theory. Here three time and space scales are required (source, observer and medium), rather than the two used in SR. Source and observer flight paths can now be compared on a universal reference (propagation medium) not relative to each other. *It is not possible to make precise comments about an observed situation unless the motion of both the source and observer are taken into account relative to the propagation medium.*

8. Despite Einstein's belief that there is no propagation medium for the propagation of EM waves, extensive evidence shows that the vacuum medium exists. The medium has finite permeability and permittivity, and because it has no mass (atomic structure) it behaves as a fluid medium. *The medium is attracted to and surrounds large gravitational bodies (planets) within its Gravitational Field of Dominance (GDOD). If the body rotates, the medium close to its surface (if gravitational) will also rotate with the body, creating a medium velocity Electro-gravitational Boundary Layer (EGBL) immediately above its surface.* The medium surrounding the planet, within its GFOD, then orbits with the planet through the medium 'stationary' within the sun's GFOD and solar system. The medium around the sun and solar system then presumably rotate with the galaxy relative to the medium basically at rest in the universe.

Chapter III

Medium Based Theories

1 Introduction

In Chapter I the basic elements of the New Relativity (NR) theory were introduced. In Chapter II, the reality of the propagation medium was considered. In this Chapter the need for the propagation medium is reviewed. Einstein's ether-less concept of relativity is considered using oblique transform axes, and compared with Lorentz's medium based rectangular transform axes theory. Measurements made in the 'early' days are investigated, Einstein's rejection of the propagation medium considered and failure to detect the medium analysed.

2 Medium Essential

The Lorentz Transform (1899) (LT) had predicted the situation before Einstein (1905) published his theory of Special Relativity (SR). The LT, before being taken up by mathematicians, was very much, and still is a physical transform, derived on solid verifiable physics. Lorentz developed his transform through propagating signals and systems moving relative to the medium. He included system time and space contraction with motion relative to the medium, the medium remaining unchanged. This transform results in the speed of light, Maxwell's Equations (ME's) (1865) differential form, and the observed events all being invariant in the moving frame. *But the propagation time in and against the direction of motion, relative to the medium, remains asymmetrical and variant, similar to the classical situation.* However, Einstein 'muddied the waters' by attempting to remove the medium, ignoring the transform's physical foundations.

Satisfying causality is critical, where the cause (source event) must always occur before the effect (observed event). Einstein's ether-less aspect of SR, not having a medium, is not a solution of the wave equation. Its observed events cannot be predicted, they can occur before the cause, as in time travel. The non measurable, non predictable aspects of SR, are based on Einstein's ether-less inertial frame. Here the propagation time is regarded as symmetrical and invariant, which is not supported by reality.

For example, sound will always drop in frequency and light red shifted, for receding sources and observers relative to the medium. This is the effect of the Propagation Time Asymmetry (PTA), it cannot happen without a medium. According to Lorentz, *EM waves behave exactly as classical waves in this respect.* They require a medium, as do Maxwell's Equations, they satisfy the wave equation, are causal and predict asymmetric propagation times up stream and down. This PTA is a basic fact for all waves moving around systems in motion. It can be obtained quite naturally by using ME's and the LT directly.

There is nothing extraordinary regarding constant or accelerative motion, Einstein believed there was a fundamental difference, he believed constant motion could not be detected. However, in this medium based theory it is the velocity at any instant that is important, whether it is constant or accelerative. Motion can be established (measured) instantly, relative to the propagation medium, either through a propagation time change, which is Galilean or through the rate of ageing, which is Lorentzian. Both of which depend on motion with respect to the medium, as Lorentz predicted. Propagation taking place directly along a line between the emission point and its reception position, without a medium (using Einstein's inertial frame), is against observed reality. Relative motion between systems is shown not to be a fundamental concept.

Further, Einstein picked and mixed aspects of the LT, i.e. he did not use the transform consistently. Generally, he elected to reject the medium, which is a vital part of the LT. But on the other hand, Einstein used Lorentz's medium based transform to derive his famous energy equation and predict other well known results. The propagation time asymmetry (PTA) and its variance, which are characteristics of the medium, are shown to be a reality in the measured data. They confirm the exact application of the LT, and therefore invalidate Einstein's ether-less aspects of SR, which are not supported by the medium based Maxwell's equations or the Lorentz transform.

3 Einstein's Ether-less Theory

It should be pointed out at the outset, that this section deals with Einstein's ether-less aspects of SR. It is an ingenious attempt to simulate the effect of no medium using medium based concepts, but it is a circular claim. Ether-less SR is therefore non causal, it cannot be measured. If one wishes to move directly to measure predictions, using a medium, this is described in Section 4. There is no physical support for the either-less simulation, apart from trying to account for the MMX, which is accountable quite naturally using a propagation medium. The ether-less theory has never been verified, and predicts outcomes that have never been measured. It is surprising that the non causal ether-less predictions have been tolerated, without proof, for over a century. Not insisting on absolute proof in research has encouraged blurring between reality and unproven speculation.

To attempt to simulate the absence of the propagation medium (ether), an oblique time and space axes transform was constructed, according to Minkowski (1908), as described by Born (1924), and later in 1962. It is illustrated in the space-time diagram in Figure 3.1 for constant and inline motion.

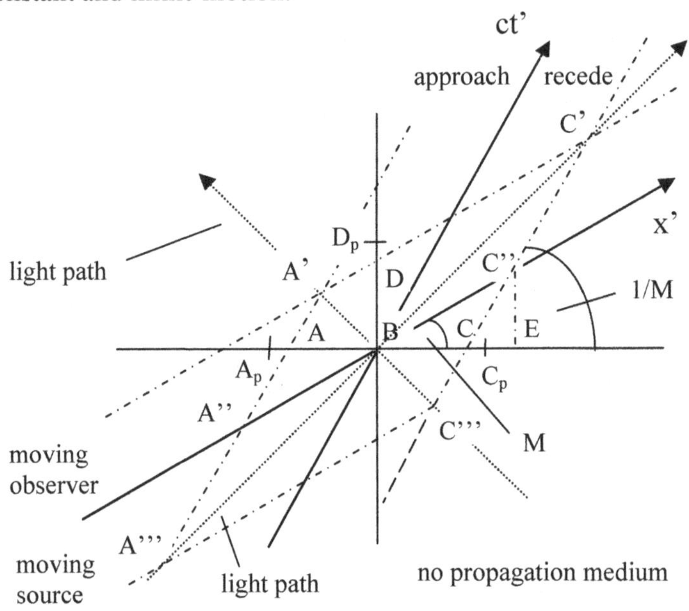

To simulate lack of a propagation medium, oblique time ct'and space x' axes are equally placed about the 45° light paths producing a symmetrical transform. The top LHS and RHS rhombuses represent observer motion approaching and receding from a stationary source at B. The bottom LHS and RHS rhombuses represent source motion approaching and receding from a stationary observer at B. (M=0.6, α=0.8).

Figure 3.1 An interpretation of Einstein's ether-less aspect of SR, simulating optical reciprocity and simultaneity

The 45° angled dotted lines BA' and BC' in the figure represent propagation light paths into and out of B in an upward direction. The 45° angle is obtained using time in years and distance in light years (LY). The relation between the oblique time and space axes ct', x', are shown symmetrically positioned each side of the light paths. The ct', x' axes have gradients 1/M and M, where M=v/c is the Mach number i.e. ratio of system speed 'v' over light speed 'c'. This makes the figure a rhombus, symmetrically placed about the 45° light paths, forming four smaller rhombuses between axes. Here all times and spaces are now equal (i.e. BA''= BC''= A''A'= C''C' etc.).

The top two rhombuses represent a stationary source at B and moving observers, approaching on the left hand side (LHS) and receding on the right hand side (RHS). The bottom two rhombuses are for a stationary observer at B and moving sources, approaching on the LHS and receding on the RHS. It can be seen from the figure that an approaching observer–stationary source time (A''A') is identical to an approaching source–stationary observer time (A'''A''), as measured on the oblique ct' axis. This makes the times equal for both source and observer motion, upholding the optical reciprocity of motion, i.e.:

A'''A''= A''A' (3.1)

This implies that the sources and observers can be interchanged or observations of each other's system time and space scales are symmetrical and interchangeable. Also, the approaching (A''A') and receding (C''C') observer times, for a stationary source are equal, satisfying simultaneity, i.e.:

A''A'= C''C' (3.2)

Thus, through a subtle rotation of the space-time axes, optical reciprocity and simultaneity are apparently simulated, giving the concept of relativity some mathematical credibility. However, there are fundamental difficulties created with this simulation. All sides (times and distances) increase with speed (rhombuses flatten and widen). Here time and space expand, for example the ratio between the BC'' in the moving frame compared with the BC_p, in the stationary propagation medium, becomes:

$$\alpha^* = BC''/BC_p = [(1+M^2)/(1-M^2)]^{1/2} = (1+M^2)^{1/2}/\alpha \quad (3.3)$$

Equation 3.3 is derived using Figure 3.1 as follows:

$$BC = \alpha BC_p \quad (i)$$
$$C''E = BE.M \quad (ii)$$

$$CE = C''E.M = BE.M^2 = (BC+CE)M^2$$
$$= BC.M^2/(1-M^2) = \alpha.BC_pM^2/\alpha^2 \quad (iii)$$

$$BE = BC+CE = (\alpha+M^2/\alpha)BC_p$$
$$= (\alpha^2+M^2)\alpha^{-1}BC_p = BC_p/\alpha, \quad \text{as } \alpha^2+M^2=1 \quad (iv)$$

Thus

$$(BC'')^2 = (BE)^2+(C''E)^2 = (BE)^2+(BE)^2M^2$$
$$= (BE)^2(1+M^2) = (BCp)^2(1+M^2)/\alpha^2 \quad (v)$$
$$BC'' = BC_p(1+M^2)^{1/2}/\alpha$$
$$= BC_p\,[(1+M^2)/(1-M^2)]^{1/2} \quad (vi)$$

Although simultaneity and reciprocity are seemingly simulated in this model, there is an expansion of $\alpha^*=(1+M^2)^{1/2}/\alpha$ in the ether-less transform simulation, compared to the well established contraction of $\alpha=(1-M^2)^{1/2}$ in the medium based Lorentz Transform (LT). It is not surprising that this oblique transform does not succeed, as it uses a medium to try and suppress the effect of a medium (circular argument). If one finds these concepts difficult to comprehend, he or she should not be too concerned as they don't represent reality. The real situation is described in the next section. There is no physical justification for this ether-less model apart from attempting to satisfy the MMX, which is justified naturally using a propagation medium.

This oblique transform representation, portrayed in Figure 3.1, is a departure from the LT. Einstein's inertial frame claim of simultaneity and invariance, in the absence of a propagation medium, is irrational (non causal). One cannot avoid the measured result predicted by the medium based LT, i.e. that the propagation time in the moving frame is variant and asymmetrical, as discussed in the next section. Although this analysis is an attempt to simulate an ether-less universe, the concepts of reciprocity and simultaniety, using an oblique axes ether-less theory, it is not a solution of the wave equation, it is only a simulation, it cannot be measured or represent reality.

4 Lorentz's Medium Based Theory

The Lorentz transform is based on a propagation medium; it predicts observer motion with respect to the propagation medium. It was developed about five years before Einstein's SR and predicts the same measured medium based results as SR. But it does not support Einstein's ether-less aspect of SR. It is shown in Section 6 and 7 that there is extensive data in support of the propagation medium, and that the causal data exhibits a variant Propagation Time Asymmetry (PTA) in the moving frame. In Section 3 it was shown that Einstein's ether-less SR, simulating Propagation Time Symmetry (PTS) and invariance, is irrational (non causal). The symmetrical ether-less space-time diagram shown in Figure 3.1 is returned back to the asymmetrical medium based diagram illustrated in Figure 3.2.

Here the ct and x axes rotate back to their Galilean rectangular positions, coinciding with the vertical and horizontal axes. These axes become the absolute time t_p in years and space x_p in light years respectively, in the real world (stationary propagating medium), legitimizing the reference for the light propagation and flight paths in the first place. The square in the figure represents the stationary source and observer case, where light from B reaches A_p and C_p simultaneously at time D_p, in the stationary propagation medium. For classical motion, the square becomes a rhombus. To accommodate the Lorentzian contraction, the classical time and space scales shrink including matter and structures by Lorentz's motional contraction factor α. The waves laid down in the medium don't change with motion. It's the moving systems space that shrink in time and space (moving frame). Thus:

$$BC/BC_p = x_o/x_p = BD/BD_p = t_o/t_p = \alpha., \text{ where } \alpha = (1-M^2)^{1/2} \quad (3.4)$$

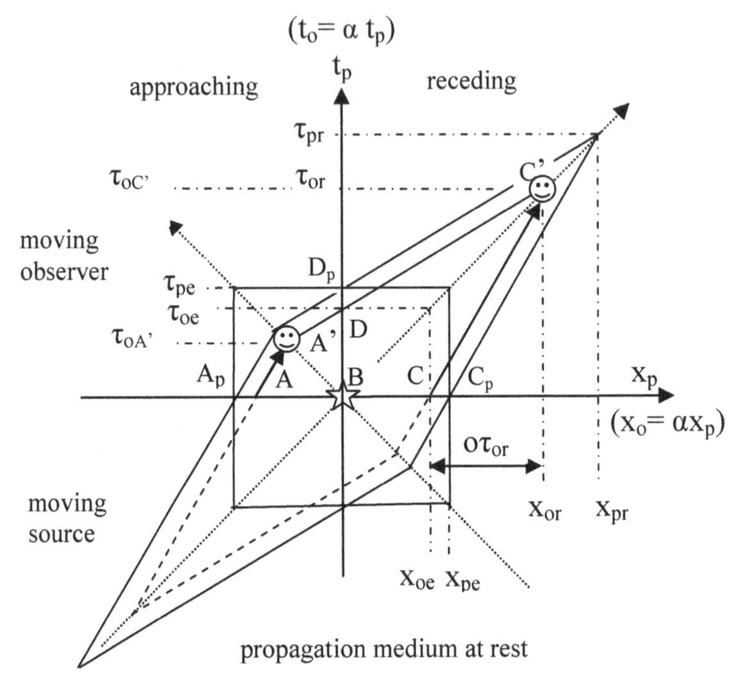

The oblique axes are returned to their rectangular positions, giving an asymmetrical Lorentz contracted Galilean Transform. AA' and CC' are now the observer flight paths with a stationary source at B. The static square becomes a classical rhombus, which shrinks by α, through relativistic motion.

Figure 3.2 In-line motion with respect to a propagation medium. The Lorentz transform is defined in the top right hand quadrant

The subscripts o and p represent relativistic observer and classical propagation paths respectively, relatively to a stationary propagation medium. Thus the outer rhombus represents classical (Galilean) motion, and the inner rhombus represents Lorentzian motion. The actual flight paths now become AA' and CC' for the approaching and receding observer respectively. Reinstating the propagating medium, it can be seen that the symmetry in Figure 3.1 is broken, invalidating the optical reciprocity of motion i.e. the viewing of each other's time and space scales in relative motion are no longer identical, equivalent or interchangeable. Also, that the propagation times up and down stream are no longer equal - simultaneity is no longer upheld, where segments of time t are now represented by τ, i.e.

$$\tau_{oA'} < \tau_{oC'} \qquad (3.5)$$

Figure 3.2 is an exact graphical representation of the LT based on the propagation medium. From this figure the LT can be identified, portrayed in the top right-hand quadrant. It can be seen that it physically corresponds to a receding observer and stationary source configuration. The other quadrants are considered in Section 2, Chapter VII. From the LT there are two motional conditions:

Condition A. Time and space contract by α relative to medium, i.e.

$$\tau_o/\tau_p = x_o/x_p = \alpha \qquad (3.6)$$

Condition B. Velocity of light 'c' is maintained in both frames, i.e.

$$c_o = x_o/\tau_o = \alpha x_p/\alpha\tau_p = x_p/\tau_p = c_p = c \qquad (3.7)$$

Distinction now has to be made between emission time 'e' and reception time 'r'. Thus the propagation times and distances, illustrated in the Figure 3.2, using equations 3.6 and 3.7 are given by:

$$x_{oe} = x_{or} - o\tau_{or} \qquad (3.8)$$

$$\alpha_o x_{pe} = c\tau_{or} - o\tau_{or} \qquad (3.9)$$

$$\alpha_o c\tau_{pe} = \tau_{or}(1-M_o)c \qquad (3.10)$$

$$\tau_{or} = \tau_{pe}\alpha_o/(1\text{-}M_o) \quad \text{or} \quad x_{or} = x_{pe}\alpha_o/(1\text{-}M_o) \quad (3.11)$$

Rearranging in mixed time and space variables

$$\tau_{pe} = \tau_{or}(1\text{-}M_o)/\alpha_o = (\tau_{or}\text{-}M_o x_{or}/c)/\alpha_o = (\tau_{or}\text{-}o x_{or}/c^2)/\alpha_o \quad (3.12)$$

Or multiplying by c

$$x_{pe}=(x_{or}\text{-}ot_{or})/\alpha_o \quad (3.13)$$

Moving Lorentzian (x_o, τ_o) and classical propagation (x_p, τ_p) frame axes are represented by (x, t) and (x', t') respectively, in the Lorentz terminology. Equation 3.11 gives the moving observer relativistic time τ_{or} in terms of stationary propagation medium time τ_{pe}, and corresponding distances. It shows the effects of the Doppler shift (1-M_o), where M_o can be positive or negative for receding and approaching observers, respectively, making the propagation time asymmetrical. It also includes the Lorentz contraction α_o, both are functions of speed with respect to the propagation medium. This Lorentzian medium based variant and asymmetric propagation time with frame motion, portrayed in Figure 3.2, is in direct conflict with Einstein's ether-less invariant and symmetrical inertial frame, illustrated in Figure 3.1.

Equation 3.12 can be recognized as the classical Lorentz Transform (LT), where the event time ratio shrinks by $\tau_{oe}/\tau_{pe}=\alpha_o$, and the propagation time $\tau_{or}=x_{or}/c$ reduces by M_o, in accordance with the Lienard-Wiechart retarded potential. Extensions to the LT, for source and observer motion in general, are portrayed in other quadrants of Figure 3.2. An approaching observer-stationary source is illustrated in the top left hand side (LHS) quadrant. A stationary observer with an approaching and receding source is represented by the lower LHS and RHS quadrants, respectively. The moving source gives a different set of equations described in graphical form, Section 2, Figure 7.1, Chapter VII and through solving the motional wave equation 5.28 and 5.31 in Chapter V.

The main differences between Einstein's oblique axes, ether-less aspect of SR, illustrated in Figure 3.1, and the rectangular axes, medium based aspect of LT, illustrated in Figure 3.2, which supports Einstein's medium based SR, are compared in Figure 3.3. The theories are based initially on Maxwell's Equations, which are based on a stationary medium. Through motion, the medium based SR (LT) retains causality, measured asymmetry, variant propagation times, and absolute time and space. Whereas, Einstein's ether-less SR, without a medium is non causal, its propagation time is symmetrical and invariant. It supports relative motion between systems, it is not supported by the medium based LT, it cannot be measured.

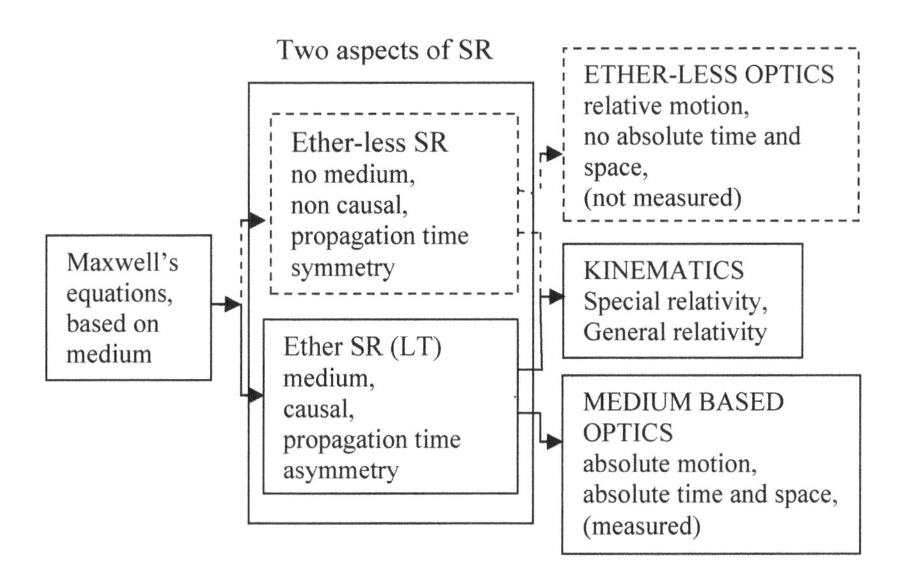

Figure 3.3 Comparison between Einstein's ether-less aspect and the medium based aspect of SR (LT)

Einstein interpreted the ether-less properties as indicating that only relative motion had meaning, leading him to derive his SR and GR, as he believed, without a medium. According to the revised theory, based on the medium based Maxwell's equations and the medium based Lorentz transform, this dotted line link cannot exist. It is impossible without the medium, making the top two right hand boxes in Figure 3.3 as inappropriate. The link must be through the medium based solid line shown at the bottom of the diagram. Thus Einstein's ether-less aspect of SR is untenable.

The Propagation Time Asymmetry (PTA) is found to be a reality in the measured data. It confirms the medium based SR and the correct application of the LT. The main impact of the medium is that absolute time and space is restored together with absolute motion with respect to the medium at rest in space. The oblique space and time axis transformation in Figure 3.1, effectively used to simulate Einstein's relativity, is then restored to the classical (Galilean) rectangular axes, Lorentzian contracted, as shown in Figure 3.2. This restores the natural asymmetry and renders invalid Einstein's symmetrical ether-less optics. Particularly, simultaneity, i.e. the forward and backward propagation times in a moving frame are no longer considered symmetrical. Also optical reciprocity, i.e. observations of each other's system time and space scales, are no longer symmetrical or interchangeable.

The re-established asymmetry restores the direction of time (causality) and defines, exactly, whether the source or observer is in motion. In spite of Einstein's rejection of the medium and his detailed description of his ether-less SR, his motional electrodynamics do in fact use a medium in solving the wave equation, otherwise his analysis would have been irrational (non causal). Einstein's medium based SR is founded on Lorentz's medium based motional theory, although Einstein gave no credit to Lorentz in his 1905 paper. *It is clear now that Einstein used a medium in his analysis, but denied its existence, causing considerable confusion for over a century.*

However, Einstein's medium based analysis is incomplete, it is described by only the top two quadrants in Figure 3.2. It cannot distinguish between source and observer motion described by all four quadrants and analytically by the solution of the wave equation, provided by EMMA, briefly described in Section 7, Chapter II and in detail in Chapter V. Thus, Einstein's SR has an ether-less aspect, which simulates simultaneity, reciprocity, no absolute time and space, and time travel, all of which are non causal and non measurable. It also has a medium based aspect that is causal, which predicts many modern day observations, but does not describe the complete process of radiation propagation and reception of sources and observers in motion.

5 Early Days

All known electromagnetic measurements and observations can be explained quite naturally using a propagation medium (ether). Historically, the medium based theories became unpopular at the end of the nineteenth century because of two primary misinterpretations of available data: Galileo's (1632) observation, which was the first to formally recognise that there was no difference between a stationary and constantly moving system (inertial frame). Besides the mechanics and the speed of light being invariant through various constant speeds, Galileo's observations, although he could not have been aware of it, could not change either. They are predicted to be invariant through the medium based solution of the optical wave equation. In equation 2.2, the motional factors $\varepsilon_s\ \alpha_s^{-1}\ \varepsilon_o^{-1}\alpha = 1$, i.e. they cancel, as described in Section 7, Chapter II.

Further, the Michelson and Morley Experiment (MMX) (1887) showed that propagation times were symmetrical and invariant on Earth moving through space. Initially, Einstein believed that the Fitzgerald (1889) spatial contraction with motion could account for the wave propagation time symmetry, but this is not so. Einstein then made the amazing claim, without proof, that effectively the mechanics, physics and electrodynamics (light propagation speed and time) were all invariant in the moving frame, implying there was no medium.

Oblique transform axes, discussed in Section 3, Chapter III, were then used to simulate the effect of no medium, attempting to simulate simultaneity (equal propagation times upstream and down (no PTA)) and reciprocity (interchanging the observer and source makes no difference optically).

However, these oblique axes predictions are not a solution of the wave equation, they are non causal. Also Einstein's assumption regarding the invariance of the propagation time in a frame moving relative to the medium is in error. According to the medium based Lorentz transform, the propagation time asymmetry (PTA) occurs upstream and down, the same as classical systems, revealing absolute frame motion with respect to the medium, as shown in Figure 1.1, illustrated in Figure 2.6 (b) and (c), and Figure 2.7 (b). Thus, although the actual mechanics, physics and observations are invariant in the moving frame, the PTA remains variant revealing motion. This is sufficient to show that the complete invariance description in the MMX is not a result of motion with respect to the medium, but a result of the medium moving with the measuring system.

If Galileo, with full sail on his galleon, had had the facility to measure the PTA upstream and down, similar to that of Sagnac, he would have found that the propagation time was variant and asymmetrical, revealing an absolute change through motion. Also, the invariance and symmetry in the MMX meant that the medium must move with the Earth, as Stokes (1845) suggested. If one accepts the presence of the stationary medium, as Sagnac's results demonstrate, there is no other rational possibility. Clearly, classical wave theory and the medium based LT (1899) supports this model, and modern data confirms it. In Galileo's case, motion is with respect to the medium creating propagation time asymmetry. In the MMX case, the medium moves with the Earth, creating symmetry in the propagation time.

This conclusion is supported by data from early and recent times: Fresnel (1818) proposed partial convection of light (increase in the velocity of light in the direction of a massive medium's motion). This is relative to a stationary mass-less vacuum medium (ether) and light source. The ability to convect light is determined by the moving medium's refractive index (speed of light in the mass-less medium compared to the speed in the massive medium). Its effect is predictable through the medium based Relativistic Addition of Velocities (RAV), discussed in Section 3, Figure 7.2, Chapter VII. *Fizeau (1851) actually confirmed partial convection of light in rapidly moving water, between a stationary source and observer, exactly according to Fresnel and RAV, as discussed in Section 4.4, Figure 7.6, Chapter VII.*

However, this convection effect is not the same mechanism as 100% light convection with gravitational induced medium movement. Here the light source, medium and observer have to move together to cause 100% light convection. Theodor des Coudres and Wilhelm Wien (1900), suggested that the degree of medium movement with gravitational bodies is proportional to gravitational mass. This would account for a range of medium motions, depending on the gravitational mass of the moving body. Stokes suggested complete gravitational induced movement of the vacuum medium around the Earth. Unfortunately, Stokes did not appear to know how to deal with the Earth's rotation and its orbital motion.

Initially, it was thought that the 'stationary' medium orbiting with the Earth should affect Bradley's (1725) star light aberration, discussed in Section 4.3, Figure 7.5, Chapter VII. Here the observer (telescope) on Earth, orbits with the 'stationary' medium surrounding the Earth, through the 'stationary' medium surrounding the Sun, and finally relative to the distant star light propagating in the medium at rest in space. This orbital motion relative to the star light causes a sideways aberration image shift of the star being viewed through the telescope. According to Fresnel (1818), a moving vacuum medium, having no mass and unity refractive index, cannot convect light, leaving the aberration unchanged.

What Stokes did not appear to anticipate is that the medium on the Earth's surface is rotating with the Earth, relative to a surrounding 'stationary' medium of finite extent, as illustrated in Figure 2.8, Chapter II. This model has been established through MMX (1887), Sagnac's (1913), and Michelson and Gale's (M&G) (1925) results. The MMX is considered in Section 2, Chapter I, and the last two references considered in detail in Section 5, Figure 4.4 and Figure 4.5 respectively, Chapter IV. Some relativists believe these three references reinforce Einstein's relativity without a medium. Again this is not possible, all three of these investigations are accounted for exactly using medium based classical PTA theory, independent of Einstein's SR. Apart from correcting automatically for Lorentzian and gravitational time slowing (long term drift), the relativistic and gravitational effect can be neglected on and around the Earth.

Einstein's contradiction of predicting ether-less properties that cannot be measured, but then using a medium in his motional electrodynamics to predict the measured results introduced an element of uncertainty, not being able to distinguish between real and imaginary situations. This confusion, between medium based and ether-less SR predictions is evident in the literature. The new theory removes the uncertainty, relying only on verifiable physics, based on a propagation medium, to re-establish reality. This medium based theory supports aspects of Einstein's medium based SR, but rejects all aspects of Einstein's ether-less SR.

6 Medium Rejection

Einstein's decision to reject the propagation medium was based mainly on the five following observations, which are correct but are explained in hind sight through the presence of the medium:

(1) The observations (and physics) do not change within a frame moving at various constant speeds.
(2) Observations of source and observer motion, moving towards each other, relative to the medium, have identical Doppler effects.
(3) The speed of light and Maxwell's equations (ME's) (1865) are unchanged in a frame moving relative to the propagation medium.
(4) The propagation time in the Michelson and Morley Experiment (MMX)(1887) is invariant and symmetrical in and against the Earth's motion.
(5) There is no evidence of a classical (mechanical) propagation medium for EM waves.

It is not surprising that the above observations were thought, initially, to indicate the absence of the medium, over a hundred years ago, when the universe was less understood. However, in hindsight these observations are explained quite naturally now using a propagation medium. In the revised theory:

Observation (1) is caused through the Doppler effect (1842), for source and observer motion, cancelling exactly, see Section 7, equation 2.3, Chapter II, where $\varepsilon_s \alpha_s^{-1} \varepsilon_o^{-1} \alpha_o = 1$ for all speeds.
Observation (2) is explained through the Doppler factors $\varepsilon_s \varepsilon_o^{-1}$ being modified by Lorentzian factors $\alpha_s^{-1} \alpha_o$ to make the effects identical, i.e. $\varepsilon_s \alpha_s^{-1} = \varepsilon_o^{-1} \alpha_o$, given by equation 2.4.
Observation (3) is explained through systems contracting through motion relative to the medium. From Section 4, equations 3.6 and 3.7, the moving frame time t_o and space x_o shrink by exactly the same fraction α with motion, compared to the stationary frame's t_p and x_p,

maintaining the speed of light c_o as invariant in the moving frame, the same as in the stationary frame c_p, i.e. $c_o=x_o/t_o=\alpha x_p/\alpha t_p=c_p$. Maxwell's Equations (ME's) (1865), resulting in the wave equation differential form, depend only on the speed of light, which therefore makes ME's invariant.

Observation (4) is accounted for through the medium moving with the Earth's surface, see Section 6, Figure 2.8, Chapter II. This is supported by Sagnac (1913), Section 5.1, Figure 4.4 and Michelson and Gale (1925), Section 5.3, Figure 4.5, Chapter IV.

Observation (5) is explained through a measured electrical fluid medium of finite permeability $\mu=1.25\text{x}10^{-6}$ N/A^2 and permittivity $ȩ=8.85\text{x}10^{-12}$ F/m, giving a finite speed of light, $c=(\mu ȩ)^{-1/2}=3\text{x}10^8$ m/s, confirming the medium presence. If there was no medium, $\mu ȩ=0$, 'c' would be infinite, which is not the case.

Thus, Einstein's Concept of Relativity (ECR) in an ether-less universe is not tenable. ECR (relative motion, no absolute time and space) is based on Einstein's Inertial Frame (EIF) (no medium, no propagation time asymmetry (PTA) or variance in the moving frame). Without a medium, ECR can give no physical explanation, for example. (a) How light disturbances propagate. (b) How the simple Doppler shift occurs. (c) How moving more dense medium than a vacuum can convect light, according to Fresnel. (d) How an impulsive super 'lightic' wave is formed in Cerenkov radiation. (e) How two systems can move apart physically, greater than the speed of light relative to the medium, but not relative to each other. These phenomena are explained quite naturally if their motions are considered relative to the medium, using the medium based Lorentz transform.

Researchers claiming the verification of the ether-less SR have usually verified results based on Lorentz's medium based transform, not the absence of the propagation medium. Without a medium, propagation time asymmetry (PTA), up stream and down, a characteristic of the propagation medium and confirmation of causality cannot be sustained. Without a medium there is no means to oppose inertial motion. These problems disappear reinstating the propagation medium. Time, being a visual effect, is predicted according to the optical solution of the wave equation, equation 5.28 and 5.31, Section 4, Figure 5.1, Chapter V.

Einstein based his SR on what he thought was an ether-less (no propagation medium) universe, where only relative motion between systems is important. However, Einstein's SR predictions can only match reality when his field equations in his motional electrodynamics, based on a medium, are used. All other ether-less predictions of Einstein's SR are non causal, they cannot, and have never been measured. Removing the medium causes insurmountable difficulties, including the inability to produce a unified theory of the universe. Einstein struggled, unsuccessfully for most of his later life, trying to achieve unification without recognising or accepting the existence of the propagation medium. However, based on his 1920 Leiden Lecture on 'Ether and Relativity' Einstein finally appeared to get close to accepting the possibility of an ether.

Einstein's measured SR predictions are founded on Lorentz's medium based transform, not on ether-less space. The ether-less predictions were supported by Born (1924) interpreted from Minkowsky's (1908) representation, using oblique transform axes, rather than the rectangular ones used in the Lorentz transform. Based on no medium, these predictions cannot be a solution of the wave equation; they are non causal and cannot be measured. A causal EM motional analysis (EMMA), based on the medium, is developed in Chapter V that extends the medium based Lorentzian transform to include both source and observer motion.

7 Medium Undetected

Now the velocity of sound in air is approximately 320m/s, relative to the speed of light of 3.10^8 m/s ($M \approx 10^{-6}$). The surface rotational velocity at the Earth's equator is of the order of that of sound, 460m/s ($M \approx 1.5x10^{-6}$). Whereas, the velocity of the Earth around the Sun is about 30 km/s ($M \approx 10^{-4}$). The Sun and the Solar System are thought to be moving in the Milky Way at around 300 km/s ($M \approx 10^{-3}$). Some nebulae are moving at thousands of km/s ($M \approx 10^{-2}$), and some distant galaxies are thought to be moving close to the speed of light ($M \approx 1$), or possibly faster. Here light emitted first would be received last (time running backwards) for an approaching source, but for a receding source there would be no dramatic observed effect, time would continue to slow down, above the light speed.

So what do these numbers mean? In this revised theory the speed of light and its propagation medium are considered to be absolute and inseparable. Therefore, the above ratios are with respect to an absolute reference (ether) in the universe. If not, how could the absolute speed of light be judged without its medium. Also, there could be no time and space continuity, no causality, the situation would be chaotic if Einstein's model was correct. Only a patchwork of discontinuous autonomous regions of relativity could exist. This is not a satisfactory situation, apparently not confronted nor reconciled in relativity. All that is required to establish continuity and causality is a medium.

So why has the propagation medium's presence been rejected for the last 100 years? What is the evidence against the ether's existence? From the 1880s to the 1930s, the ether believers including MM (1887), Miller (1925), Michelson, Pease and Pearson (1929) and Kennedy and Thorndike (1932) tried extremely hard to detect the effect of the motion of the Earth through the ether. They measured the round trip propagation time (RTPT) by using more and more elaborate interferometers. However, the expected time change was not measured. By the late 1930s, these measurements were finally interpreted as a negative result. The effect was for the establishment to reject the ether and support Einstein's Concept of Relativity.

In retrospect this should not have happened. From the evidence available, the establishment could just as well have concluded that the ether was stationary with the Earth, as Stokes (1845) suggested and Sagnac (1913) demonstrated. Sagnac showed with his rotating mirrors, that the medium existed and was stationary on the Earth's surface. Here motion relative to the Earth's surface gave variant propagation time asymmetry. This PTA was in accordance with the medium based LT, equation 1.2, demonstrated in Figure 1.1. It is analysed graphically in Figure 3.2 and expressed algebraically by equation 3.11. Also Michelson and Gale (1925) provided further evidence that not only did the medium exist, but it moved with the Earth's surface relative to a 'stationary' medium surrounding the Earth as discussed in Section 6 Figure 2.8, Chapter II. Amazingly, this evidence was not acted upon. Einstein ignored, or even might not have been aware of either of these pioneering medium confirmations.

After the Second World War, precision increased using modern measuring standards such as Mossbauer Fe^{57}sources (X-rays exciting iron radiation levels), with uncertainty (variability) less than 3.10^{-13}. Using more accurate standards, further methods have been used to investigate the Earth's motion relative to the ether, such as reflecting transmission lines, r f interferometers, tuned systems such as resonant cavities and lasers. All of these supported the MMX interferometer results. They were all in vain, giving an invariant symmetrical propagation time, confirming Sagnac's findings that the medium moved with the Earth's surface.

It was thought perhaps that one way propagation time (OWPT) measurements might detect the motion of the Earth through the medium. This OWPT method is much more sensitive to speed than the

RTPT, where the propagation time delays in each direction tend to cancel. According to Section 5 Chapter VI, the RTPT$\approx M^2b/c\alpha$ and OWPT$\approx 2Mb/c\alpha$ where b is the source - observer distance and $\alpha\approx 1$ for typical speeds on Earth ($M=10^{-6}$). It can be seen through the M^2 term that the RTPT is not very practical for typical speeds on and around the Earth ($M^2=10^{-12}$). If b=20km and the Earth's velocity≈30km/s ($M\approx 10^{-4}$) orbiting around the Sun, then it is shown that the OWPT≈ $(2b/c)M/\alpha\approx 4x10^4x3^{-1}x10^{-8}\ x10^{-4}=12x10^{-9}$=12ns, which is well within modern measurement accuracy.

Krisher et al (1990) made such measurements in a fibre optic cable in each direction, simultaneously. They used phase comparison of an ultra stable f=100 MHz hydrogen maser. This gave a theoretical number of 2π phase reversals (fringes) N=OWPTxf=$12X10^{-9}x10^8$=1.2, but none were measured. The measurements were made over several rotations of the Earth, but could only measure a velocity corresponding to 0.1km/s, rather than the Earth's orbital motion of 30 km/s. Again, implying there is little relative motion, i.e. the medium moves with the Earth, confirming Sagnac's result. However, the fibre optic cable moving with the Earth and maser, compromised the result (no relative motion). It would have been better to have used micro waves moving relative to the vacuum (open air), rather than use a fibre optic cable.

A OWPT measurement in a frame moving on Earth was attempted. Unfortunately the experimenters made their measurements normal to the direction of motion, along an arm rotating relative to the Earth and medium. Turner and Hill (1964) and later Riis et al (1988), with greater precision, used this approach but declared their results negative, 'confirming' the absence of the ether. However, in this case, there would be negligible detectable one-way propagation time change broadside to the direction of motion, thus invalidating their conclusions. If they had made the measurements in the direction of motion, as in Sagnac's rotating mirrors, they would have measured a propagating time asymmetry, predicted according to Lorentz, Section 4, equation 3.11.

Newman et al (1978) increased the confidence of the non-ether believers by listing verified SR experiments ‘with an even higher degree of accuracy’. Again, the crucial point was missed that it was the medium based aspects of LT that was being verified, not the ether-less SR or the medium’s absence. The above RTPT and OWPT, measurements, are based on detecting the motion of the Earth through the ether. The fact that none of these methods have detected any appreciable motion does not confirm the absence of the propagation medium. It confirms that there is negligible motion between the Earth’s surface and the propagation medium rotating with it, as originally confirmed by Sagnac (1913).

8 Conclusions

8.1 Main concepts

1. ***Propagation time asymmetry.*** All wave theories, both classical and EM, predict wave Propagation Time Asymmetry (PTA) around moving sources relative the propagation medium. PTA is a confirmation of causality, where the source event (cause) must always occur before the observed event (effect).

2. ***Lorentz fundamental.*** The Lorentz transform is the algebraic solution of the EM motional wave equation based on a propagation medium. The solution includes the essential classical PTA, and an additional modifying Lorentzian time rate and system contraction (LC) through high speed motion relative to the medium. At Earth speeds and short periods of time, LC is negligible compared to PTA, generally the classical instantaneous PTA dominates on Earth.

3. ***Einstein’s ether-less SR.*** Einstein assumed there was no medium (ether) needed to propagate light (observed events) through space.

Without its medium, the EM wave equation cannot be solved, without its solution measured events cannot be predicted. This ether-less theory, based on Einstein's inertial frame, cannot support the medium based PTA. It lacks the requirement of a medium, it is non causal.

4. ***Einstein's non causal predictions.*** Einstein's ether-less claims are: i) only relative motion between systems can be detected, ii) distinction cannot be made between stationary and constantly moving frames, iii) the medium is redundant. These claims have never been measured; they are non causal (false). Einstein's ether-less inertial frame, responsible for these beliefs, having no medium, is inappropriate to predict measured SR observations. It's Lorentz's medium based optical frame that correctly predicts the events.

5. ***Einstein's measured predictions.*** Although Einstein claimed ether-less predictions, his field equations are founded on the medium based LT. The solution of his EM wave equation is therefore causal, predicting measured observations. The same predictions can be obtained directly from the medium based classical wave equation, modified by the medium based LC. Therefore, Einstein's measured predictions are medium based, similar to any other classical propagating wave theory, not ether-less as he believed.

6. ***Incomplete theory.*** Einstein's medium based theory is incomplete; it does not distinguish between measured source and observer motional differences, or between stationary and moving medium reference frames. A general EM Motional Analysis (EMMA) is developed, which accounts for measured source and observer motional effects relative to the medium, not possible without a medium. This is achieved using three sets of time and space scales, rather than the two used in SR

8.2 Distinguishing features, SR and NR

1. ***Non causal waves.*** Einstein believed EM waves were special, they did not require a propagation medium (ether) to transmit light. Einstein offered no alternative wave propagation mechanism. Medium

removal is against well established physical principles. Without a medium the wave equation cannot be formed or solved resulting in non causal predictions that cannot be measured.

2. ***Relative motion.*** Without a medium, Einstein claimed that observations depend only on relative motion between systems, predicting non causal time travel and no absolute time and space, neither of which can be measured. Whereas, NR based on a medium is causal, it establishes that all motional theory and observed effects are based on motion with respect to the medium.

3. ***Two aspects.*** Einstein's SR, has two aspects: An ether-less aspect based on no medium, which is irrational (non causal), its predictions cannot be measured. And a medium based aspect based on Lorentz's medium based motional transform that does predict causal observations that can be measured.

4. ***Motion distinction.*** However, Einstein's medium based predicttions do not distinguish between system and media motion. NR extends Lorentz's medium based transform to predict the important differences between sources and observers in motion, and between stationary and moving media.

5. ***Medium entrainment.*** Einstein claimed that space was uniformly empty. Whereas NR not only establishes that the medium exists over all space, but it is not homogeneous, it is attracted to and compressed around large gravitational bodies.

8.3 Characteristics of NR

1. ***Vacuum medium.*** All waves require a propagation medium to propagate. The only difference between classical and a vacuum medium is that time and structure shrink passing through the EM medium and gravity compresses the medium both time and space.

2. ***Medium finite.*** It is futile to deny the EM propagation medium's existence. It has well established, measured transmission capabilities, electrical permeability (inertia) and electrical permittivity (stiffness) to bounce the wave disturbances through the medium.

3. ***Medium removal.*** It is against basic physics to attempt to remove the propagation medium. There is no authority or justification to remove the medium. Without a medium, the wave equation cannot be formed or solved.

4. ***Medium absence.*** In an ether-less universe there is no known way of transmitting EM waves, creating inertia, propagation time has no direction, distinction cannot be made between which system is moving, the source or observer, which system is ageing least, or identify the optical paths - they are indeterminate.

5. ***Medium interaction***. All motional effects, including classical and relativistic, (time slowing, systems contraction, propagation time asymmetry and invariant speed of light) are caused through interaction with the medium, as Lorentz predicted, not through relative motion between systems, as Einstein believed.

6. ***Medium theories.*** All rational (predicable) wave theories are medium based, including the EM wave equation, Maxwell's equations, Lorentz's transform and Relativistic addition of velocities, they are all meaningless (non causal) without their propagation medium.

7. ***Medium link.*** The medium is a common link between system motion, accelerating frames and gravity. The medium fills the space part of space-time, and space-time distortion is the compression of the medium's space and time, by gravity, according to Schwarzschild.

8. ***Medium fluidity.*** The mass-less medium (vacuum), having no atomic structure, behaves as a compressible electric fluid, which is attracted to and moves with large gravitational bodies. If the body rotates (and there is a gravitational atmosphere), the medium close to the body's surface will tend to rotate.

9. ***Medium common.*** The same propagation medium appears to transmit all electric fields, including EM, gravitational and inertial. A common medium determines the same wave propagation speed for all disturbances, i.e. photons, gravitons and bosons, respectively.

10. ***Speed of light.*** The measured vacuum medium properties are finite giving a finite propagation speed, no medium would give infinite speed. The speed is invariant only because both time and space shrink

by exactly the same ratio in the moving frame, maintaining the propagation speed.

11. ***Light convection.*** A moving heavy medium (containing atoms and molecules) is shown by Fizeau to drag (partially convect) light propagating in a vacuum medium, predicted according to Fresnel, and the medium based Relativistic Addition of Velocities (RAV).

12. ***Rectangular axes.*** Minkowski's rectangular axes four vector space-time analysis is medium based. The vertical and horizontal axes of the space-time diagram use Lorentz's rectangular axes medium based transform, representing time and space respectively. There is no mechanism to remove the medium.

13. ***Oblique axes.*** These axes are a simulation, they simulate simultaneity and reciprocity, neither of which can been measured. This model attempts to remove the effect of the medium by using medium based concepts (circular argument). It is not a solution of the wave equation, it is non causal, it does not represent reality.

14. ***Types of frame****.* Einstein's ether-less invariant inertial frame cannot support Propagation Time Asymmetry (PTA). Having no medium, Einstein's inertial frame is non causal, it is incapable of transmitting observations. Whereas, Lorentz's optical frame, containing the medium and its waves, facilitates the observations.

15. ***Types of motion.*** One is where systems move with respect to the medium (causing PTA). The other is where there is no motion (no PTA) between the frame and medium, as in the medium moving with gravitational bodies.

16. ***Motional angles.*** In GPS, **source** (satellite) motion relative to the medium causes a backward displacement angle with the normal to the direction of motion. In Bradley's stellar aberration, **observer** (telescope) motion relative to the medium causes a forward angle relative to the star's actual position.

17. ***Time travel.*** Causally it is possible to visually visit the past, but not to interact with it. It is not possible to visit the future, it is non causal, it has not yet occurred. However, it is possible to slow one's

time down by moving at high speed or visiting a heavy gravitational body, but this is not reversible time travel.

18. ***Hybrid frames.*** The speed of light can be exceeded across hybrid reference frames, $M^*>1$, where $M^*=M/\alpha$, $M=v/c<1$, where M is the Mach number, system speed 'v' in the medium compared to the light speed 'c', and α is the Lorentz contraction factor, allowing almost limitless distances in space travel to be achieved.

19. ***LT, Gravity and acceleration.*** LT contracts systems moving through the medium, gravity compresses the medium and observer motion relatively expands the medium. Gravity can be reduced (neutralised) through observer acceleration, giving an optical explanation of the time Equivalence Principle.

20. ***Gravity and inertial field.*** Gravity is the steady difference electric field from dissimilar charges in atoms and molecules. Inertial field is the residual gravity from all the matter in the universe.

Chapter IV:

Media and Frames

1 Introduction

In this Chapter classical and EM media are compared showing those classical motional aspects that are common to both, and the additional relativistic effects that become important at high speed. The effect of systems and frames moving relative to the medium are considered, distinguishing between Einstein's Inertial Frame (EIF) and the Optical Propagation Frame (OPF) that transmits the observations. Also, distinction is made between source and observer motion relative to the medium, with particular reference to source backward and observer forward motional angles. Experiments describing motion on and above the Earth for both classical and relativistic motional effects are described. These include Sagnac (1913), Michelson & Gale (1925) and Hefele & Keating (1972), three very important historic measurements, establishing the EM medium's presence.

2 Classical and EM Media

2.1 Historic development

To predict observations, through electromagnetic waves (light), the correct physical approach should have been to set up and solve the EM wave equation with respect to its propagation medium, the same as for classical theories. Huygens (1690) introduced the concept of a wave theory of light relative to a propagation medium (ether). Young (1804) much later, demonstrated the interference of light waves relative to the medium. Fresnel (1818) theoretically, and Fizeau (1851) practically, showed that EM waves could be partially convected by a moving dense medium relative to a vacuum. Maxwell (1865) derived his EM field equations using Huygens wave model and Faraday's (1844) electrical theory, both medium based. Finally, Hertz (1888) showed that EM waves and light were the same thing. These pioneers set the framework for the natural development of the EM motional wave equation and its solution, based on a propagation medium.

2.2 Medium discarded

Unfortunately, this propagation medium development did not happen. The medium approach was abandoned as the Michelson and Morley Experiment (MMX) (1887) showed that the Round Trip Propagation Time (RTPT) was invariant, i.e. there was no observed effect on the optical waves moving across and in the direction of the Earth's motion, either through rotation, orbiting around the Sun or moving with the galaxy, relative to the medium at 'rest' in space. The invariance in the MMX cast doubt on the existence of the medium. Further, investigations could not detect any evidence of a physical ether either. As a result, Einstein (1905), encouraged by Minkowski (1908), discarded the medium all together. At a stroke, Einstein effectively dispensed with absolute time and space and replaced it with relative motion. This apparent lack of an ether, discredited any subsequent attempts to develop a motional EM wave equation solution for source and observer motion relative to the medium. This has lasted right up to the present time, the beginning of the twenty-first century.

2.3 Classical medium

A deformable medium is essential to support a field that transmits compression waves. A field is a physical quantity that propagates disturbances in its steady state value, in predictable time order, propagating spatially between two points. For example, in acoustics the medium is the air, the steady field is provided by gas molecules, providing the atmospheric pressure P, and its perturbation pressure (sound) ΔP, is predicted by Navier-Stokes fluid dynamic equations. The waves are generated by an acoustic source, for example a loudspeaker of volume displacement strength q. The transmitted sound energy is $(\Delta P)^2/z$, where the acoustic impedance to wave motion is given by z=ρc. Here ρ is the density of air and $c=(dP/d\rho)^{1/2}= (\gamma P/\rho)^{1/2}$ is the velocity of sound, γ is the ratio of specific heats and γP is a compressibility term. Thus the acoustic wave compressions ΔP (bouncing through the medium) depend on the classical medium's density ρ and springiness γP. In addition to the propagation mechanism, a classical (mechanical) medium has viscosity that causes drag (opposes the source motion through it), even for constant velocity, it also attenuates the wave with propagation distance, through atmospheric absorption mechanisms, etc.

2.4 Electrical medium

All types of wave propagation require an equivalent mass/spring system to bounce the waves through its propagation medium. The electrical medium that transmits light, or photons, is no exception. Its properties are well defined and measureable. It is characterized by an equivalent permeability μ (an electrical inertia or inductance term), and a permittivity ȩ (or ε) (an electrical stiffness or capacitance term). In the case of optical media, μ varies little between materials, it's ȩ that determines the light properties. A diamond has a large ȩ, varying widely with frequency, giving it its spectacular sparkle. Unlike classical wave propagation, light can be polarised, but its medium has no viscosity or attenuation terms, apart from spherical spreading.

In an analogous fashion to acoustics, where the undisturbed air pressure P provides the sound field, the steady electric field E provides the field in electromagnetic theory. Its equivalent perturbation to sound ΔP is ΔE, accompanied by a magnetic perturbation ΔH, generated by an EM source of charge strength q. To describe EM waves, Navier-Stokes Fluid Dynamic Equations (NSE's) are replaced by Maxwell's Equations (ME's), discussed in Chapter V. Changes in q, either strength or spatial positioning, produce wave disturbances ΔE and ΔH resulting in light energy transmission of $(\Delta E)^2/z$. The q variations could be lightning, electrons jumping orbit in atoms, or moving at high speed in an accelerator. In a similar form to air the electric medium's equivalent wave impedance is z=μc, where $c=dE/d\mu=(1/\mu ȩ)^{1/2}$ is the speed of light. In an exactly equivalent way to sound, the EM wave speed depends on the electrical medium's inertia μ and springiness ȩ (or ε). It can be seen that basically there is little difference between EM waves and classical waves, they both require a medium and have a finite propagation speed.

In an unstressed vacuum (free space), $\mu=1.25\text{x}10^{-6}$ N/A^2 and ȩ=8.85x 10^{-12} F/m, giving $c=(\mu ȩ)^{-1/2}=3\text{x}10^8$ m/s. For gases, including a vacuum, ȩ is small (electrically soft), up to 100 times softer than solids. Space without a medium, μ=ȩ=0, would give an infinite propagation speed, which is not the case. Therefore a vacuum is not empty space. With no mass (no rigid atomic structure), a vacuum behaves as a compressible electrical fluid medium. It appears to be attracted to and forms medium gradients around moving gravitational bodies, planets etc., including the Earth. Denying the presence of the well documented propagation medium is not only irrational, but it is against fundamental physics, based on elementary principles of wave theory. There is nothing ether-less (non causal) regarding EM wave propagation.

Maxwell's Equations (ME's) predict energy transmission and information (source events) to the far field, at the speed of light, relative to its medium. No other mechanism is required, there is no mechanical ether for EM waves. ME's provide the basic EM wave equation solution, equivalent to NSE's, discussed in Section 2.3, predicting electric and coupled magnetic disturbances propagating along the electric field lines. ME's have a characteristic wave propagation speed and time delay (retarded time) between the source emission and the observer reception positions. These are 'static' equations, describing stationary systems in a stationary propagation medium.

3 Moving Frames

3.1 Lorentz transform

The Lorentz Transform (LT) (1899), was first suggested by Fitzgerald (1889) for system contraction moving with respect to the propagation medium. Larmor (1897) proposed time and structure contraction in the case of the electron, and the systematic form of the LT was developed by Poincaré (1900) based on a propagation medium. The solution of the classical wave equation, for moving systems, is based on the Galilean transform between stationary and moving reference frames. This results in a varying propagation time delay between a source and observer of fixed separation, moving with respect to the medium.

The only major difference between classical and EM waves, is that there is an equal fractional shrinking of time rate and space with motion, according to Lorentz. This makes the speed of light and ME's equations invariant in the moving frame, whereas in classical systems the propagation speed remains variant. The transform is often treated as a purely mathematical tool. Inspection reminds us that it is indeed based on a physical medium. This is the physics of EM wave motion, it does not support Einstein's invariant inertial frame, based on no medium. Lorentz was dismayed at the use of his transform in Einstein's relativity, apparently without its medium. Lorentz never accepted the medium's removal and resisted it to his death.

3.2 Einstein's inertial frame

Mechanically, Einstein's Inertial Frame (EIF) is one moving in constant rectilinear motion (constant momentum), where Galileo (1632), first observed there was no change (invariant mechanics) through motion. Basically this was Einstein's first postulate. He used this fact, plus he assumed optically that EIF would create no Propagation Time Asymmetry (PTA) between a source and observer of fixed separation, stationary or in motion. Without proof or providing any mechanism, it was assumed that the propagation time, in the x direction of motion (t_x=x/c) and in the y direction, perpendicular to the motion (t_y=y/c), was invariant (unchanged through motion), creating Propagation Time Symmetry (PTS). The situation is illustrated in Figure 4.1(a), where it is assumed that the propagation speed 'c' in the stationary and moving frames, and in both the x and y directions, is independent of motion.

This invariance of the speed of light realised through Lorentz's transform, is Einstein's second postulate. Mainly, as a result of the Michelson and Morley Experiment (MMX)(1887), which appeared to support this model, it was assumed that there was no difference between a stationary and moving frame. This was interpreted as rendering the medium redundant. However, in Section 6, Figure 2.8, Chapter II, it is shown that there is no relative motion between the Earth and its medium. The MMX is therefore not a frame moving relatively to the propagation medium, and as such will have no motional properties. Einstein's ether-less inertial frame having PTS and not PTA has never been measured, it is non causal.

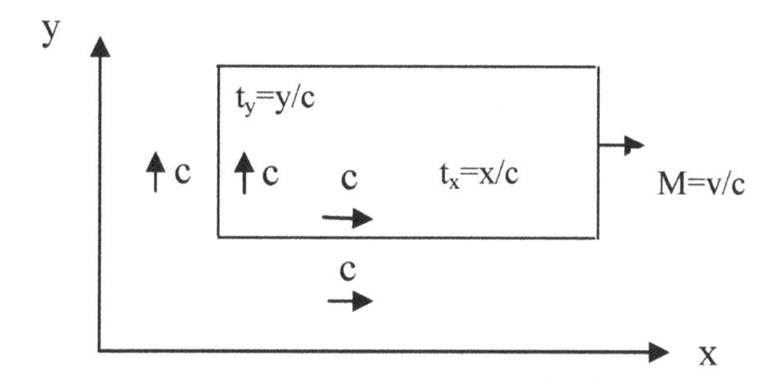

(a) Einstein's Inertial Frame (EIF), no propagation medium - non causal, Propagation Time Symmetry

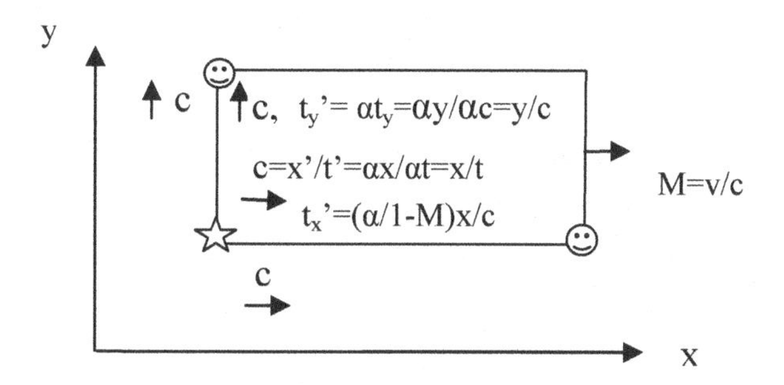

(b) Lorentz's Propagation Frame (LPF), propagation medium - causal, Propagation Time Asymmetry

Figure 4.1 The difference between Einstein's Inertial Frame (EIF) and Lorentz's Propagation Frame (LPF), only perpendicular to motion does LPR support EIF

3.3 Optical propagation frame

For motion relative to the EM medium an Lorentz's Propagation Frame (LPF), depicted in Figure 4.1(b), is used. Observations require wave propagation, and therefore a propagation medium, either moving with the frame as in the MMX, or stationary in the universe. In this latter case, observations between a fixed source and observer moving with respect to the stationary medium are shown, where the propagation speed is now with respect to the stationary medium.

According to the LT, time contracts (time slows) ($t'=\alpha t$) and space contracts ($x'=\alpha x$) in the moving frame, in the x direction, by same fraction i.e.

$$c=x'/t'=\alpha x/\alpha t=x/t. \quad (4.1)$$

This maintains the propagation speed 'c' in the x direction, in the moving frame, the same as in the stationary propagation medium, considered further in Section 3, equation 5.19, Figure 5.3, Chapter V.

Regarding propagation in the y direction, perpendicular to frame motion, here the propagation speed is again c, but not with respect to some unknown propagation mechanism moving with the frame. The propagation is again relative to the stationary medium, but in this case the propagation has to be at an angle with the direction of motion to be received by the observer, as considered in Section 4.1, Figure 7.3 Chapter VII. This results in an extended propagation distance y/α, and increased propagation time $t_y=y/\alpha c$ in the medium, where α is a purely geometrical term, which happens to be exactly the same as the Lorentz contraction factor. Thus the increase in propagation time in the medium, is exactly offset by the Lorentz time contraction (slowing) α, in the moving frame., i.e.

$$t_y'= \alpha t_y=\alpha(y/\alpha c)=y/c \quad (4.2)$$

This maintains the effective propagation time t_y' as invariant perpendicular to the direction of motion. Thus, although the propagation speed is actually c in the stationary medium, through a propagation geometric quirk, the effective propagation speed remains c in both the x and y directions in the moving frame complying with EIF. *However, unlike the y direction, the propagation distance and time in the x direction does change through frame motion relative to the propagation medium, making the frame variant.* This is similar to classical theory, not in accord with EIF, it is given by equation 3.11, Section 4, Chapter III, as:

$$t_x' = [\alpha/(1-M)]x/c, \quad M = v/c \quad (4.3)$$

Here M can be positive or negative, in and against the direction of motion, making the propagation time asymmetric and variant. With no propagation medium, EIF cannot distinguish between a fixed source and observer stationary or moving at constant velocity, it can have no motional characteristics. Whereas, the PTA in the Lorentz's Propagation Frame (LPF) gives sources and observers their directional features with motion, relative to the medium, according to the LT. Sagnac confirmed this PTA and variance with motion relative to the propagation medium, as early as 1913, described in detail in Section 5.1, Figure 4.4, Chapter IV. This important experiment, supporting the LT and its medium, went unheeded. There was no hard evidence or authority for Einstein to assume there was no medium, whereas Sagnac's experiment confirmed the medium's existence precisely.

3.4 EIF Inappropriate

Thus the LT does not support Einstein's Concept of Relativity (ECR) based on Einstein's Inertial Frame (EIF), where there is considered to be no medium and no PTA in the moving frame. The disturbance propagation time (retarded time), resulting directly from the classical Galilean transform, is variant in the moving frame, both in the direction and perpendicular to the direction of motion, making the motion absolute compared with the stationary medium. *The inclusion of Lorentz's frame contraction, in the direction of motion, only makes the propagation time invariant perpendicular to the direction of motion.* It does not remove the Galilean variance, in the direction of motion. Although the LT is in agreement with Einstein's first and second postulate (physics and speed of light are invariant), it does not support EIF, where the essential observations (propagation times) are excluded through the lack of a medium. The PTA and variance is shown clearly to be a reality in the data confirming the exact application of the variant LT, invalidating the invariant EIF and ECR.

3.5 Oblique transform axes

Based on his Invariant Inertial Frame (EIF), Einstein declared there is no medium and that only relative motion is meaningful, inferring no absolute time and space, thus creating his concept of relativity. To attempt to explain the removal of the propagation medium, Minkowski's geometry was used as discussed in detail in Section 3, Chapter III. Here the rectangular time and space axes, used by Lorentz, were substituted for oblique axes to remove (simulate the non-existence, but not actually eliminate) the medium. This simulation is a departure from the LT and therefore is not supported by it. The introduction of the oblique axes, leading to ECR, attempted to justify:

1. Simultaneity - equal propagation times for equal distances upstream and down (propagation against and with motion).
2. Reciprocity - moving frames look alike from each others reference (interchanging source and observer makes no difference optically).
3. No absolute reference - an infinite number of inertial frames moving at different speeds are considered to exist (all inertial frames are identical).
4. No propagation medium - indicates no absolute time and space supporting time travel.

Although this manoeuvre seems to simulate the above characteristics, it appears they cannot match reality. Artificially removing the effects of the propagation medium is irrational. The ether-less predictions are not a solution of the wave equation, they are non causal, they cannot be measured. The argument is circular, the medium is attempted to be removed based on medium concepts.

Thus, the crucial difference between Einstein's ether-less theory and one with a propagation medium, in accordance with the LT, is the measured asymmetry and variance of the propagation time. The MMX observations are explained now quite naturally with respect to the propagation medium rotating and orbiting with the Earth. Contrary to popular belief, the MMX's invariance (no PTA) does not disprove the existence of the medium. Nor does it support Einstein's ether-less relativity. The medium moves with the Earth satisfying basic wave theory.

4 Source and Observer Motion

4.1 Collision in space

Relativists consider that there is no difference between a constantly moving and a stationary reference frame. Also that interchanging the source and observer makes no difference to the observations. In this revised theory, it is shown that although the observations in the moving frame are invariant, the PTA changes with motion revealing motion. Interchanging the source and observer, discussed in Section 4.3, Chapter VI, also changes the observations for the same flight paths through the medium, again in disagreement with Einstein's Inertial Frame (EIF) predictions. Relativists claim that there is no difference out in space if a lamp-post hit one's space-car or the car hits the lamp-post, as illustrated in Figure 4.2, they believe in only relative motion.

Mechanically they are right. However, when it comes to observing these events, particularly at high speed, the observed event will be quite different depending on whether the vehicle or lamp-post moves with respect to the propagation medium. It involves the whole optical processes of radiation, propagation and reception with respect to the medium. Such complete processes are absent in Einstein's ether-less invariant inertial frame and therefore not available to establish the effects of motion.

Figure 4.2 According to Einstein's relativity, it is not clear what caused the collision, the space-car or the lamp-post. According to the revised theory, the one that moved relative to medium at rest in space caused the damage

On Earth it's obvious who causes the damage; one has the Earth and its propagation medium as a reference. But even in space it is still clear; the one who moves with respect to the medium, at rest in space, causing the PTA, will be seen by the insurance company to have caused the damage. Motional characteristics of a moving system, relative to the medium, will become obvious at high speed, through considerable wave compression in front of the system in motion, as witnessed by Cerenkov radiation, discussed in Section 1 and illustrated in Figure 2.2(b), Chapter II. Relativists will find this hard to accept, if only the speed of light were lower this would become very clear (visibly obvious to all).

4.2 Moving observer

Evidence supporting the propagation medium has been available for many years, but has not been taken seriously since the development of Einstein's Special Relativity (SR). Bradley (1725), Figure 7.5, showed classically that an observer (telescope) moves relative to a star and medium at rest in the universe. He noticed that a telescope moving with the Earth in orbit around the Sun (rather than its rotational velocity, which is hundred times slower than the orbital velocity), has a forward (aberration) angle slightly in front of the star to be viewed.

The angle reverses with the Earth's direction twice a year as the Earth orbits the Sun. The aberration is a necessary alignment to allow the light waves, propagating through the stationary medium in 'space', to reach the bottom of the moving telescope and into the eyepiece. The concept is similar to tilting an umbrella to keep one's feet dry when walking in the rain. This is a fundamental demonstration of the Earth (observer) moving relative to a medium at rest in space. The verification of the aberration of telescopes using Bradley's argument and including the full kinematic (relativistic) addition of velocities, using a propagation medium, is considered in Section 4.3, Chapter VII.

Stellar aberration is a first-order classical effect requiring a medium. There is no way to avoid the role the propagation medium plays in the telescope tilt process. Relativists have tried to dismiss the medium with relativistic arguments. However, the aberration of telescopes cannot be explained, at Earth speeds, through using a weak sixth order relativistic effect, it is a first order classical (Galilean) property of the propagation medium.

The aberration is not caused through the resolved angle between the Earth's orbital motion and the speed of light, assumed with no medium. The angle is actually embedded in the 'stationary' medium surrounding and orbiting with the Earth. The angle is not affected by adding water to the telescope, demonstrating the medium's presence. The medium provides a smooth transition between the medium moving with the Earth and the stationary one in space. Bradley showed that the medium is fixed, effectively stationary with respect to the star and the universe, and that the observer (telescope) moves with the Earth relative to the stationary medium in space.

4.3 Relativistic addition of velocities

In this revised theory motion can only occur between a moving observer and source relative to the medium, not directly with each other. Relativistic Addition of Velocities (RAV), based on the LT is some times attempted to be used to support Einstein's relativistic motion between systems, based on no medium. But this is not possible as RAV itself is a medium based concept. One cannot use arguments based on the medium to prove there is no medium. RAV is actually based on the LT and Lorentz's medium, the effect is not dependant on relative motion between systems, but on Lorentzian system motion relative to the medium. At relatively low speeds, experienced on and around the Earth, the relativistic effect (Lorentz contraction) is small compared to the classical PTA and can usually be neglected. The relativistic effect is not considered again until Section 5.6.

4.4 Moving source

Brecher (1977), 250 years after Bradley's discovery of aberration, deduced, amongst others, that the light propagation speed from rapidly orbiting binary stars approaching and receding from the Earth, is relative to a stationary medium at rest in space, not moving with the orbiting star. Figure 4.3 illustrates three possibilities. Figure (a) depicts light behaving like projectiles without a medium. Here, the propagation speed adds to the source speed, creating gaps and overlapping radiation (light from the approaching path overtaking the slower radiation from the receding path). This is not Doppler shifting; it is light distortion, which is not what is measured. Light does not have projectile like properties. This is not to be mistaken with mass-less light waves considered as a burst of energy impulses (photons) propagating at the speed of light with respect to the propagation medium.

One possibility, which does not produce overlapping radiation, through source direction reversal, is shown in Figure (b). Here the medium exists over all space and moves instantaneously with the source. This would require infinite field reaction speed over all space and again the propagation speed would add to the source speed, which again is not what is measured in the far field. The only possibility, matching observations, is illustrated in Figure (c). Here the propagation medium is effectively at rest in space and the star moves relative to it. The propagation speed is constant with respect to the stationary medium causing Doppler shifted frequency, not overlapping radiation. The additional effect of the medium moving locally, with the star, is discussed in Section 3.3, Figure 8.3, Chapter VIII.

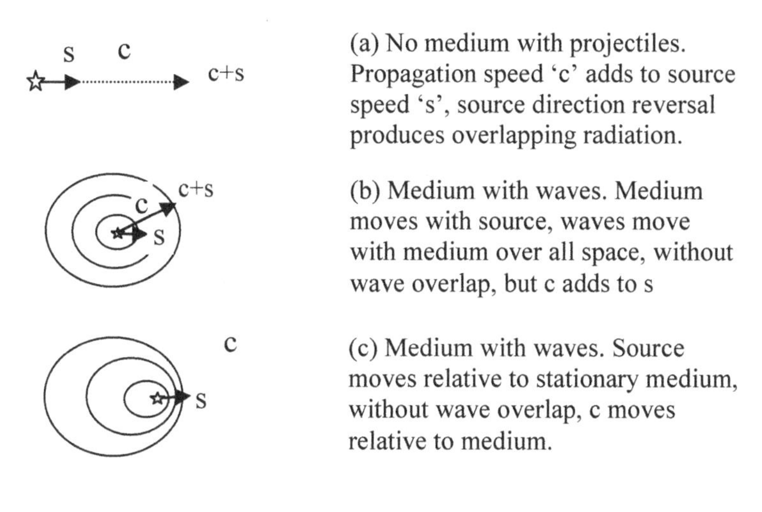

Figure 4.3 Source motion with three propagation possibilities. (a) No medium with projectiles. (b) Medium moves with source. (c) Source moves relative to medium

4.5 Moving source angle

A telescope has no aberration (apart from its own motion) when viewing a rapidly moving star. It aligns on the earlier source emission position, within the stationary propagating medium, at the later reception time. During this time the source has moved on. It is similar to observing and listening to a fast moving aircraft. One has to look in front of where the sound is coming from, to see the plane. A source (satellite) moving with respect to the medium creates a backward motional angle, as illustrated in Figure 8.5, Chapter VIII. The angle is formed in Global Positioning Systems (GPS) (1992), it is a result of waves, from a moving satellite (source), propagating at an effective angle in the stationary medium surrounding the Earth.

The signal propagates between the satellite and Earth station, which rotate laterally to this signal propagating in the stationary medium forming the displacement angle. These details, available since the early 1700s for observer motion and more recently for source motion, confirmed that both sources and observers move relative to the medium. If the existence of the medium is doubted, then the Bradley (observer) and satellite (source) observations should be attempted to be accounted for without a medium, or medium based concepts. These motional descriptions are based purely on classical (non relativistic) arguments.

4.6 Absolute motion

Evidence for absolute acceleration relative to the residual gravity field in the universe was discussed in Section 3, Chapter II. Absolute motion, using simple classical methods, has been established relative to the universe, the medium and the Cosmic Microwave Background (CMB). This radiation discovered by Penzias and Wilson (1965), radiates uniformly and omni-directional in the propagation medium throughout space. Absolute motion has been confirmed through the Cosmic Background Explorer (COBE) (1992), measuring the increase in collective wave energy in the direction of motion relative to the CMB and the medium.

The increase in energy manifests itself as a directivity increase in the received radiation superimposed on the universe microwave back-ground variations. This is a fundamental classical measurement of the motion relative to the universe, and the propagation medium. The measurement does not rely on time and space change, propagation distance change nor spectral shifting/event time changes to establish the effect of motion. It is an absolute measurement of motion with respect to the omni-directional radiation propagating in the propagation medium. Thus motion can always be measured, through Propagation Time Asymmetry (PTA) relative to the vacuum medium or relative to the CMB radiation both assumed, on average, to be at rest in the universe.

5 Motional Theories

There are two time changing effects with motion. One through classical PTA and the other through relativistic contraction. The classical PTA is considered first and then the relativistic effect.

5.1 Sagnac's rotating mirrors

Away from gravitational bodies, the EM radiation propagates relative to the stationary medium at rest in free space. On a large gravitational body (planet), light propagation is shown to occur in the medium close to a body's surface moving with the body. Sagnac (1913), a long time ago, established this fact. He found that the medium existed and moved with the Earth's surface. Sagnac measured propagation time asymmetry (PTA) with motion in and against the stationary medium. He passed light around a rotating system of mirrors, as illustrated in Figure 4.4 (a).

Light moving in the direction of the mirror motion was delayed passing around the rotating optical loop in the 'stationary' medium, arriving at the observer/detector at a later time $t=d/c$. Here d is the optical path length around the rotating system and 'c' is the propagation speed in the medium. In the mean time, the rotating system moves on, spatially by Δd, given by the system peripheral speed 'v' multiplied by the time delay. Passing the light around in the opposite direction to the mirror motion reduces the propagation delay. Therefore we have Δd=vt=vd/c=Md, where M=±v/c, causing fringe movement in an interferometer. The corresponding incremental propagation time delay then results:

$$\Delta t=\Delta d/c=vd/c^2=Md/c=Mt=2^{1/2}2\omega A/c^2 \qquad (4.4)$$

Here v=ωr, ω is the angular speed, r is the radius, t=d/c, d=4l, $l=2r/2^{1/2}$ and $A=l^2=2r^2$. As the light is propagated around a loop, back to its starting position the propagation is one-way. This enables the propagation delay between the moving source and observer to be directly dependent on speed M, rather than M^2 as in the MMX, described in Section 6.2, Chapter I, which is less sensitive to speed.

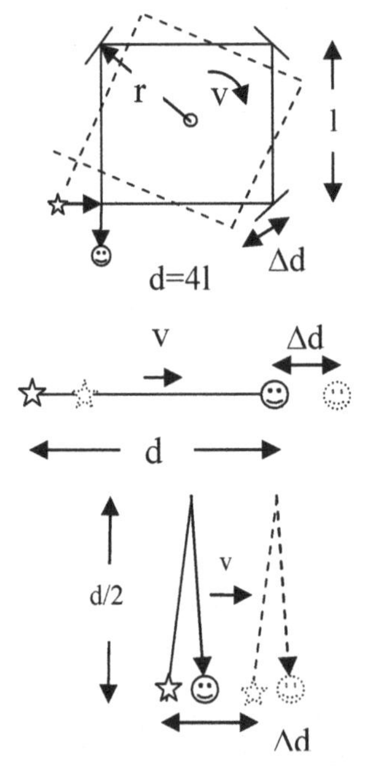

Light passes around a mirror system in a stationary propagation medium of peripheral distance d in a propagation time t=d/c. If mirror system moves at velocity v then the incremental distance travelled in time t is Δd=vt=vd/c.

(a) Rotating system

Similarly the propagation time t in stationary medium between two fixed points of separation distance d is t=d/c. If linear system moves at velocity v then the incremental distance travelled in time t is Δd=vt=vd/c $\Delta d/c=vd/c^2=Mt$. The distance d can be normal to the direction of motion

(b) Linear speedometer

Figure 4.4 Sagnac's rotating mirrors generate Propagation Time Asymmetry (PTA) delay through motion relative to the medium

The phase difference and the number of interference fringes N (number of 2π phase rotations) is then given by:

$$\Delta\varphi=2\pi f\Delta t \text{ and } \Delta t=\Delta d/c=vd/c^2=Mt$$

$$N=\Delta\varphi/2\pi=f\Delta t=fvd/c^2=fMd/c=Md/\lambda \quad (4.5)$$

Where f is the frequency and λ is the wavelength of the source (laser). If v=300m/s ($c=3x10^8$, $M=10^{-6}$), $\lambda=6x10^{-7}$m and d=3m, then $N=Md/\lambda$ $=10^{-6}x3x6^{-1}x10^7=5$ fringes. Whereas for the MMX, $N=M^2d/\lambda=10^{-12}x3$ $x6^{-1}x10^7=5x10^{-6}$ fringes, i.e. 10^{-6} smaller.

5.2 Sagnac established medium

Sagnac's method provided for the first time an absolute demonstration of the existence of the medium and motion with respect to the medium. If the medium had not existed, or moved with the moving system as in the MMX, then there would have been no effect, giving no time delay and no fringe movement. The detection of the Propagation Time Asymmetry (PTA) was damning evidence. It established beyond doubt motion with respect to the medium, confirming the exact application of the LT with its PTA. This confirmed the invalidation of EIF and ECR, with its symmetrical propagation time, discussed in Section 3.4.

If the medium was doubted, these important measurements could have been pursued further. For example, they could have been performed on a vibrating or rotating arm, described in Sections 4.4 and 5.2, Chapter VI. Or using a one way propagation configuration on a fast moving train. Unfortunately, none of these methods were attempted. There seemed to be no interest to challenge relativity, particularly as Einstein was occupied with his newly released GR, in 1915. Sagnac's experiment had sufficient accuracy to confidently demonstrate, for the first time, the existence of the medium stationary with the Earth's surface, accounting exactly for MMX null result, where no effect of the Earth's motion could be found on light propagation on Earth. However this absolute demonstration of motion relative to the medium appears to have been completely ignored.

5.3 Michelson and Gale

Whereas, Sagnac's rotating mirror system measured motion relative to the Earth's surface and its medium, Michelson and Gale (M&G) (1925) used a large Sagnac optical loop, stationary on the Earth's surface. This configuration depicted in Figure 4.5 is capable of measuring propagation on the Earth's surface relative to the 'stationary' propagation medium surrounding the Earth. The experiment used loop area as the main parameter. However, it can be explained more fundamentally using a rectangular circuit with its long sides latitudinally separated. In this case the latitudinal arms would have slightly different velocities at slightly different distances away from the Earth's equator, due to the Earth's rotation. The latitudinal arms then determine the velocity, whereas the longitudinal arms, being perpendicular to the direction of motion, would have no velocity differential, but would determine sensitivity.

5.4 Medium boundary layer

Sagnac has shown on the Earth's surface there is no relative motion between the Earth and its medium (100% convection). Therefore the medium must rotate with the Earth. In this case propagation media on the Earth will move with different surface speeds v_1, and v_2, depending on their distance from the equator, of surface speed v_e, relative to the 'stationary' medium surrounding the Earth, as shown in Figure 4.5 (a). The propagation time along each latitudinal arm of equal length d is:

$$t_2=d/(c+v_2) \text{ and } t_1=d/(c+v_1) \quad (4.6)$$

The time difference is then:

$$\Delta t = t_2 - t_1 = d[\{1/(c+v_2)\} - \{1/(c+v_1)\}]$$

$$= d[(c+v_1-c-v_2)/(c+v_2)(c+v_1)] \approx d\,(v_1-v_2)/c^2 \qquad (4.7)$$

For v_1 and $v_2 << c$

$$v_1 - v_2 = v_e(\cos\varphi_1 - \cos\varphi_2) = 2v_e \sin[(\varphi_1+\varphi_2)/2]\sin[(\varphi_1-\varphi_2)/2] \approx 2v_e(h/2R)\sin\varphi \qquad (4.8)$$

where $\sin(\varphi_1+\varphi_2)/2 \approx \sin\varphi$, and $\sin(\varphi_1-\varphi_2)/2 \approx \sin(h/2R) \approx h/2R$

Equation 4.7, using equation 4.8, then becomes:

$$\Delta t \approx (d/c^2)2v_e(h/2R)\sin\varphi \approx (dhv_e/Rc^2)\sin\varphi \approx (A\omega/c^2)\sin\varphi \qquad (4.9)$$

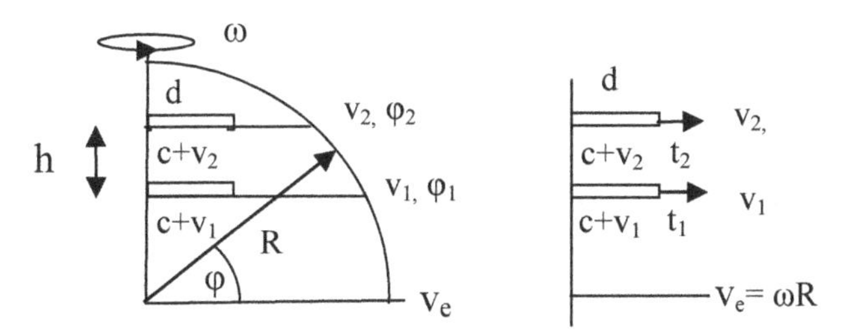

(a) Propagation with respect to medium moving with the Earth

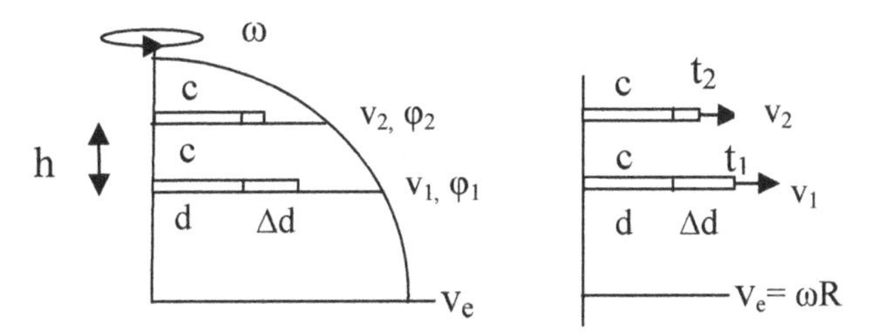

(b) Propagation with respect to stationary medium surrounding the Earth

Figure 4.5 Michelson and Gale's fixed optical loop on Earth's surface measures surface propagation speed relative to a 'stationary' medium surrounding the Earth

This result, where $A=dh$ is the loop area, $\omega=v_e/R$ and φ is the subtended angle to the equator, was demonstrated by M&G using an interferometer to measure the time difference. If the medium, in contact with the Earth, is not rotating with the Earth, but is considered stationary with the surrounding medium, then the situation is as illustrated in Figure 4.5(b). Here incremental distance relative to the stationary medium would be given by $\Delta d=vt=vd/c$, giving the time as:

$$t = (d+\Delta d)/c=d(1+v/c)/c \quad (4.10)$$

The difference in the propagation times is then:

$$\begin{aligned}\Delta t &= t_2-t_1=d[\{(1+v_2/c)/c\}-\{(1+v_1/c)/c\}] \\ &= d[\{(c+v_2)/c^2\}-\{(c+v_1)/c^2\}]=d(v_2-v_1)/c^2\end{aligned} \quad (4.11)$$

As $v_2<v_1$, v_2-v_1 is now negative, making $t_1>t_2$ and Δt negative.

Measurements show that $t_1<t_2$, confirming that the contact medium is in fact rotating with the Earth's surface and not stationary with the surrounding medium. Thus there must be a medium boundary layer between the moving surface and the surrounding 'stationary' medium. If there was no stationary medium surrounding the Earth, then $v_e \rightarrow v_o \pm v_e$ i.e. v_e in equation 4.9 would include the Earth's orbital speed v_o, plus the asymmetry term $\pm v_e$. As the Earth's orbital velocity $v_o \approx 100$ times that of its rotational velocity v_e, the effect would be obvious. This is not measured, therefore the surrounding medium must move with the Earth in its orbit around the Sun, and not with respect to the medium at rest in space (Solar System).

5.5 Medium confirmation ignored

M&G's measurement was an important result, it demonstrated that there was relative medium motion between any two latitudinal positions on Earth. This indicates that the propagation medium is effectively 'clinging' to the Earth's surface. That is, the medium is not only moving, but moving at different speeds according to its position from the equator, with the lowest velocity (zero) on the axis of rotation (poles). M&G's measurements confirmed Sagnac's result, that the propagation medium exists and moves with the Earth's surface.

Additionally, M&G had identified the existence of the medium gradients, created by the curvature and rotation of the Earth, and invented a method of measuring them. However, more importantly, M&G's results imply that the Earth's rotation and the medium in contact with its surface rotates relative to a 'stationary' medium surrounding and moving with the Earth, i.e. a medium boundary layer is formed between the rotating surface and the surrounding 'stationary' medium orbiting with the Earth.

Thus (i) the presence of the propagation medium, (ii) motion with respect to the medium, (iii) surface medium velocity gradient, (iv) a boundary layer above the surface and (v) a 'stationary' medium surrounding and moving with the Earth, could have all been inferred in the 1920s using only first order classical methods. *These medium evidences appear to have been completely overlooked by Einstein and his contemporaries, a scientific oversight of considerable magnitude.* This did not help in the acceptance of the medium, which has been avoided in the literature right up to the present time.

5.6 Relativistic time

The above methods are based purely on Galilean classical wave dynamics which have been adequate to measure motion on Earth, i.e. Lorentian time and space contraction have not been included. Using more elaborate equipment, Picasso et al (1977), for example, again confirmed motion with respect to a medium (stationary on Earth), only this time using relativistic (Lorentzian) time change. They showed that a Muon's time τ_s, rotating at 99.94 per cent of the speed of light around a 14 metre diameter storage ring at CERN, Switzerland, runs much slower (half life extended by 30 times) compared to the Earth's time τ_p. According to equations 5.27, and 5.30, Chapter V, for a source moving at velocity 's' relative to the stationary medium and stationary observer, o=0, $\alpha_o=1$, the source time τ_s becomes:

$$\tau_s=\alpha_s\tau_p,\ \ \alpha_s=(1-M_s^2)^{1/2},\ \ M_s=s/c, \quad (4.12)$$

for M<<1

$$\alpha\approx(1-M^2/2),\ \ \Delta\alpha\approx-M^2/2 \text{ and } \Delta\tau_s\approx\Delta\alpha\tau_p\approx-M^2\tau_p/2 \quad (4.13)$$

As an example, for the Earth rotating at the equator with respect to its surrounding medium, where s=460m/s, $M=s/c=460/3x10^8\approx1.5x10^{-6}$ $\Delta\alpha\approx-M^2/2\approx-1.5^2x10^{-12}/2\approx-1.1x10^{-12}$, and τ_p=24 hours=$8.6x10^4$s, thus $\Delta\tau_s\approx\Delta\alpha\tau_p\approx-1.1x10^{-12}x8.6x10^4$=-95ns/day i.e. the Earth's time at the equator, through motion, is running ≈ 95ns/day slower than its poles.

5.7 Asymmetric ageing

Further, Hafele and Keating (H&K) (1972) found that the relativistical time of an aircraft slowed more rapidly flying with the rotation of the Earth than against it. The asymmetry results through the aircraft's time being relative to the 'stationary' time surrounding the Earth τ_p, rather than with respect to the Earth's time τ_e, illustrated in Figure 4.6.

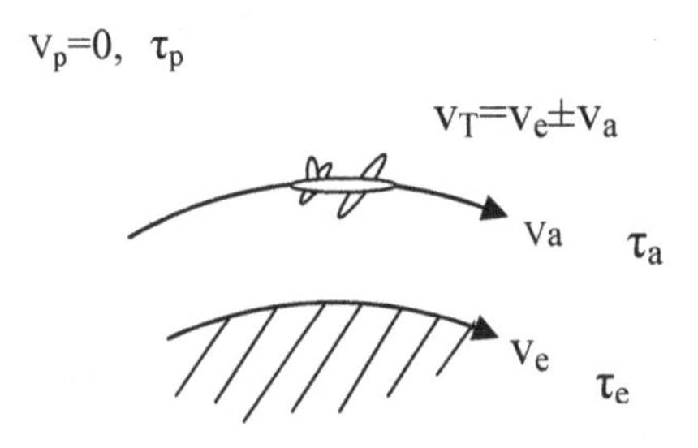

Figure 4.6 Aircraft time τ_a and Earth time τ_e relative to the surrounding 'stationary' propagation medium time τ_p

Here the aircraft's time τ_a slowing depends on its speed and the Earths rotational speed compared to the stationary medium surrounding the Earth

Let the aircraft's time τ_a compared to the stationary medium time τ_p be:

$$\Delta\tau_a \approx \Delta\alpha_a \tau_p, \text{ where } \Delta\alpha_a \approx -(v_e \pm v_a)^2/2c^2 \qquad 4.14)$$

and the Earth's time τ_e be:

$$\Delta\tau_e \approx \Delta\alpha_e \tau_p, \text{ where } \Delta\alpha_e \approx -v_e^2/2c^2 \qquad (4.15)$$

The difference time between the aircraft and the Earth's time is then:

$$\Delta\tau_{a\text{-}e} = \Delta\tau_a - \Delta\tau_e \approx -(\Delta\alpha_a - \Delta\alpha_e)\tau_p \approx -\tau_p [(v_e \pm v_a)^2 - v_e^2]/2c^2 \approx -\tau_p(v_a^2 \pm 2v_e v_a)/2c^2 \qquad (4.16)$$

where $\tau_p \approx \tau_e$, as $\tau_p - \tau_e$ is very small. If $v_a \approx v_e$, from equations 4.14, and 415, time change relative to stationary medium is:

$$\Delta\tau_e \approx -\tau_p M_e^2/2, \ \Delta\tau_a \approx -\tau_p 4 M_e^2/2 \text{ and } 0 \qquad (4.17)$$

And from equation 4.16, relative to Earth going West and East respectively:

$$\Delta\tau_{a\text{-}e} \approx +\tau_p M_e^2/2 \text{ and } \approx -3\tau_p M_e^2/2, \quad M_e = v_e/c \qquad (4.18)$$

For simplicity, if $v_a = v_e = 460$m/s ($M \approx 1.5\text{x}10^{-6}$) (Earth's velocity at the equator), then flying a resurrected Concorde for example, for $\tau_p = 24$ hours ($8.6\text{x}10^4$s) around the equator, supersonically (>320m/s) at v=460m/s, then the time change relative to the stationary medium will be $\Delta\tau_e$ =-95ns/day on Earth, and $\Delta\tau_a$ =-380ns/day and 0ns/day on the aircraft in and against the Earth's rotation. With respect to the Earth's time, the aircraft's time will be, $\Delta\tau_{a\text{-}e} \approx$-285ns/day and +95ns/day.

Hafele and Keating, actually flew around the Earth but at a slower speed and at an altitude of approximately 10km, close to the equator. Correcting for small gravitational time changes, values of 59ns and -273ns were measured relative to the Earth's time, compared with calculated values of 40ns and -275ns, for West and East flight directions, respectively, actual speeds were not given. *The aircraft's time change corresponds to time slowing with motion relative to the 'stationary' medium surrounding the Earth, not relative to the Earth's motion.* If there were no surrounding 'stationary' medium, these times would become asymmetrical based on the Earth's rotation relative to its orbital motion, which is a much larger asymmetry, but not measured.

6 Conclusions

The only motional differences between classical and EM media are that time and structures contract at high speed passing through the EM medium. Classical Propagation Time Asymmetry (PTA) effects are common to both media. Clear distinction is made between Einstein's Inertial Frame, which without a propagation medium cannot predict observed motional properties, and the Optical Propagation Frame (OPF) with a medium that predicts measured motional properties. Distinction is made between source and observer motion with respect to the propagation medium. Bradley's (1725) stellar aberration (observer motion relative to the medium) causes a forward angle displacement in front of the star to be viewed. Whereas, source motion with respect to the medium causes a backward angle established in GPS (1992) and satellite communications. It has been confirmed classically by Sagnac (1913) that the medium exists and rotates with the surface of the Earth. Also it has been confirmed by M&G (1925) classically, and H&K (1972) relativistically, that a medium surface boundary layer (less than 10km) exists between the moving surface medium and 'stationary' medium surrounding the Earth.

Chapter V:

Motion and Gravity

1 Introduction

In this chapter, the EM Motional Analysis (EMMA) is developed from solving the EM wave equation, for sources and observers in motion with respect to the propagation medium. The solution can be treated as two distinct motional processes, a classical and relativistic one. The basic transform is classical with rectangular time and space transform axes. This transform describes the wave generation, propagation and reception between sources and observers in motion relative to the propagation medium. The second transform is relativistic, it describes the additional Lorentzian shrinking of the classical time and space in a frame moving relative to the medium.

These transforms are expressed in three time and space scales, the source event time and space t_s x_s, the observer event time and space t_o x_o and the propagation medium event time and space t_p x_p. The three time and space scales allow any source/observer situation to be analysed. Schwarschild's (1916) medium compression through gravity, both time and space, is also included. Source motion relative to the medium contacts time and space and observer motion relatively expands the medium both time and space. Thus observer motion provides the ability of reducing the effect of gravity, explaining a time Equivalence Principle.

2 Classical (Galilean)Time

When one thinks of motion, one naturally thinks in terms of motion relative to the Earth or stars etc. Arbitrary motion between moving systems, without a reference, is an abstraction we Earthlings have difficulty with. The natural reference for observation is the propagation medium that transmits them. Away from large gravitational masses, the medium is shown to exist and be at rest over all space, unambiguously defining whether the observer or source or both move relative to the medium. This applies generally for all wave forms. It's in accord with the classical transform, between a stationary and moving frame. For EM systems, there is the additional contraction (Lorentz transform) of time and spatial structure, through motion.

In the case of acoustic sources, the solution is based on Navier-Stokes Equations (NSE's) of fluid motion. It is clear here that both the source and observer move relative to its propagating medium (air). In the case of electromagnetic sources, the wave motion is now based on Maxwell's electromagnetic field equations. The predicted disturbances have a wavelength, propagating speed and Doppler shift. These are tell-tale characteristics of waves propagating relative to a medium. The similarity of these electromagnetic wave forms with acoustic waves was the motivation to pursue an electromagnetic wave equation solution relative to its electromagnetic medium.

Because of the relatively slow motion on Earth, compared to the speed of light, what is seen is pretty well what is actually happening. However, because of the huge speeds and distances within the universe, one's view of these events can be affected. To compare these changes one can assume time to be invariant and the disturbance to change. This is appropriate for simple harmonic functions where the frequency can be considered to change (frequency domain). However, for more complex functions it is more convenient to consider the disturbance to be invariant and the time to change (time domain).

2.1 Stationary wave equation

For those readers who are not familiar with wave theory it is probably more useful to move directly to the graphical representation in Section 2.3. The propagation medium for the transmission of EM waves was considered in Section 2.4, Chapter IV. Here the medium has electrical properties of permeability μ and permittivity ę (or ε), EM impedance z=μc and propagation speed $c=(\mu ę)^{-1/2}$. For an EM disturbance propagating relative to the medium, Maxwell's four electro-magnetic field equations are:

$$\text{div } D = \rho \quad \text{Gauss electrical} \quad \text{(i)}$$

$$\text{div } B = 0 \quad \text{Gauss magnetic} \quad \text{(ii)}$$

$$\text{curl } E = - \partial B/\partial t \quad \text{Faraday inductive} \quad \text{(iii)} \qquad (5.1)$$

$$\text{curl } H = \partial D/\partial t + J \quad \text{Ampere-Maxwell} \quad \text{(iv)}$$

In the above equations, D=ęE, where D is the electrical displacement and E is the electric field. B=μH where B is the magnetic flux density and H is the magnetic field strength. ρ is the electric charge density and J is the current density. In a vacuum, where there is no charge or current, ρ=0, J=0, $\nabla \cdot D = 0$, then the last two equations above reduce to:

$$\nabla \times E = -\partial B/\partial t, \qquad \nabla \times B = \mu ę \, \partial E/\partial t, \qquad (5.2)$$

Here we have used the shorthand symbols div ≡ ($\nabla\cdot$) and curl ≡ (∇x). If an electromagnetic source singularity of charge strength Q(t) is introduced at the origin of the coordinate system, then one can write equation 5.2 as:

$$\nabla \times E = -\partial B/\partial t, \qquad \nabla \times B = \mu \, ę \, \partial E/\partial t + Q(t) \qquad (5.3)$$

Again $1/\mu\varsigma=c^2$ corresponds to the propagating speed squared, i.e. (speed of light)^2 relative to the propagation medium and its electric field. Taking the curl of both sides of the first equation in 5.3 one has:

$$\nabla \times (\nabla \times E) = \nabla \times (-\partial B/\partial t) \qquad (5.4)$$

Using the standard result $\nabla \times (\nabla \times A)=\nabla(\nabla \cdot A)-\nabla^2 A$, the left-hand side (LHS) of equation 5.4 becomes:

$$\nabla \times (\nabla \times E) = \nabla(\nabla \cdot E) - \nabla^2 E = -\nabla^2 E \qquad (5.5)$$

as $\nabla(\nabla \cdot E)=0$. Inserting RHS of equation 5.5, into the LHS of equation 5.4 and inserting RHS equation 5.3 one has:

$$-\nabla^2 E = \nabla \times (-\partial B/\partial t) = -\partial/\partial t(\nabla \times B) = -c^{-2}\,\partial^2 E/\partial t^2 - \partial Q(t)/\partial t, \quad (5.6)$$

Thus the 'stationary' electromagnetic wave equation becomes:

$$\nabla^2 E - c^{-2}\partial^2 E/\partial t^2 = \partial Q(t)/\partial t,$$

Using further shorthand symbols the wave equation can be written as:

$$\Box^2 E[t] = \partial Q[t]/\partial t, \quad \text{where} \quad \Box^2 = \nabla^2 - c^{-2}\,\partial^2/\partial t^2 \qquad (5.7)$$

For a simple one dimensional solution in the x direction, $\nabla^2=\partial^2/\partial x^2$, generally, the outward solution for the second order differential equation 5.7 becomes:

$$E[\tau] = (4\pi R)^{-1}\partial Q[\tau]/\partial t, \quad \text{where} \quad \tau = t - R/c \qquad (5.8)$$

t is the familiar universal (observer) time and τ is the source emission event time. R is the disturbance propagation distance between the source emission and the observer reception position in the medium. τ is sometimes called the retarded time because it can be expressed in terms of the observer time t and the propagation delay time R/c between the source and observer.

2.2 Effect of motion

The general solution of the wave equation for classical sources and observers in motion with respect to its propagation medium is available, Wright (1986). The static EM wave equation 5.7 can be motionally transformed as follows. Let the source and observer move at velocity 's' and 'o', respectively, relative to the propagating medium. Let the source and observer times be represented by t_s and t_o, respectively, and t_p be the time in the stationary propagating medium space x_p. Following the format of equation 5.7, the motional wave equation can be written as:

$$\Box_p^2 (\Delta_o E[t_p]) = \partial(\Delta_s Q[t_p])/\partial t_p, \quad \Box_p^2 = \nabla_p^2 - c^{-2}\,\partial^2/\partial t_p^2 \qquad (5.9)$$

where the local motion delta function Δ defines the effect of observer and source motion on their local time compared to the medium time as:

$$\Delta_o E[t_p] = \int_{d\to o} \Delta_o(x-(ot_p-nd))\, E[t_p]\, dx = E[t_o],$$

$$t_o = \varepsilon_o^{-1} t_p, \quad \varepsilon_o = (1-M_o\cos\sigma_o), \quad M_o = o/c \qquad (5.10)$$

$$\Delta_s Q[t_p] = \int_{d\to o} \Delta_s(x-(st_p-nd))\, Q[t_p]\, dx = Q[t_s],$$

$$t_s = \varepsilon_s^{-1} t_p, \quad \varepsilon_s = (1-M_s\cos\sigma_s), \quad M_s = s/c \qquad (5.11)$$

Δ_o and Δ_s are instantaneous point moving (local motion) operators describing sources and observers moving repeatable over a distance d in the propagating medium. The operators have the properties of describing motion at a static point in the medium. As the flight path d goes to zero, n goes to infinity, transforming time from that of a stationary point to motion at a point. Further, σ_s and σ_o are the angles between the individual source and observer motion directions and the propagation path between the source emission and the observer reception positions, as illustrated in Figure 5.1. M_s and M_o (light Mach numbers) are the source s and observer o velocities compared to the wave propagating speed c, and ε_s and ε_o are the dynamic motional operators describing the effect of motion on source and observer time, respectively.

Thus the stationary EM wave equation 5.7 can be extended for both source and observer motion relative to the propagation medium, based on the classical transform, as:

$$\square_p^2 E[t_o] = \partial(Q[t_s])/\partial t_p, \qquad \square_p^2 = \nabla_p^2 - c^{-2}\,\partial^2/\partial t_p^2 \quad (5.12)$$

Its solution, equivalent to equation 5.8, then becomes:

$$E[\tau_o] = (4\pi R)^{-1}\partial Q[\tau_s]/\partial t_p, \quad \tau_o = t_o - R/c \qquad 5.13)$$

Where the time relationships in equations 5.10 and 5.11 now become:

$$\tau_p = \varepsilon_o\tau_o, = \varepsilon_s\tau_s, \quad \text{or} \quad \tau_o = \varepsilon_s\,\varepsilon_o^{-1}\,\tau_s, \quad (5.14)$$

Equation 5.14 gives the relationship between the source event time τ_s and observer event time τ_o for motion relative to a stationary propagating medium of time τ_p. The situation is illustrated in Figure 5.1, showing arbitrary source and observer flight paths, where R_i is the initial separation distance, R_o is the initial propagation distance and $R^\dagger$ is the instantaneous source/observer propagation distance between the source and observer. t_o is the observer time which is equal to the observer event time τ_o plus the propagation time R_o /c. For a source and observer moving together $\varepsilon_s\,\varepsilon_o^{-1}=1$, the observed event time does not change with motion, $\tau_s=\tau_o$. At low speeds an approaching source and observer give approximately the same effect $\varepsilon_s \approx \varepsilon_o^{-1}$.

Event and propagation distance time changes with motion are our main interest here. However, there are other motional effects which are not considered. For example, the theory is for a point (compact) source where the wavelength is large compared to the source size. For non-compact sources, changes in the wavelength with motion will occur, affecting the source radiation impedance and observer reception interference. There is also the concentration and dilution of the radiation through source and observer motion relative to the medium.

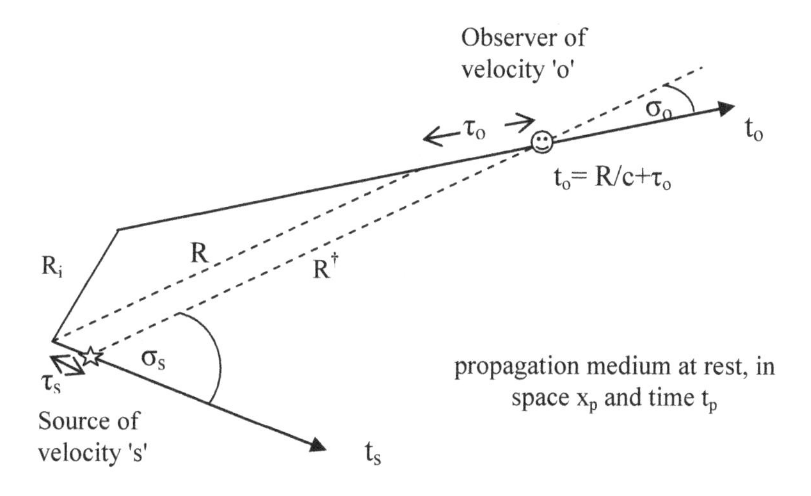

Figure 5.1. Source and observer flight paths, not in the same direction, moving with respect to a stationary propagation medium

2.3 Graphical representation

Figure 5.2 illustrates the classical time changes, graphically, in a space-time diagram, for source and observer motion, again with velocities s and o respectively (gradients in the diagram). For simplicity, the motion is in the same straight line, $\sigma_s = \sigma_o = 0$. The bottom left-hand side of the figure shows the source motion with its initial source-observer separation, propagation and continually changing propagation distances R_i, R and $R^{\dagger}$ respectively. These produce the corresponding propagation times by dividing by c. During the source event time τ_s the source travels a distance $s.\tau_s$. The receding observer motion is shown at the top right-hand side of the figure. During the observer event time τ_o, the observer travels a distance $o.\tau_o$ to the right. From Figure 5.2 it can be seen that:

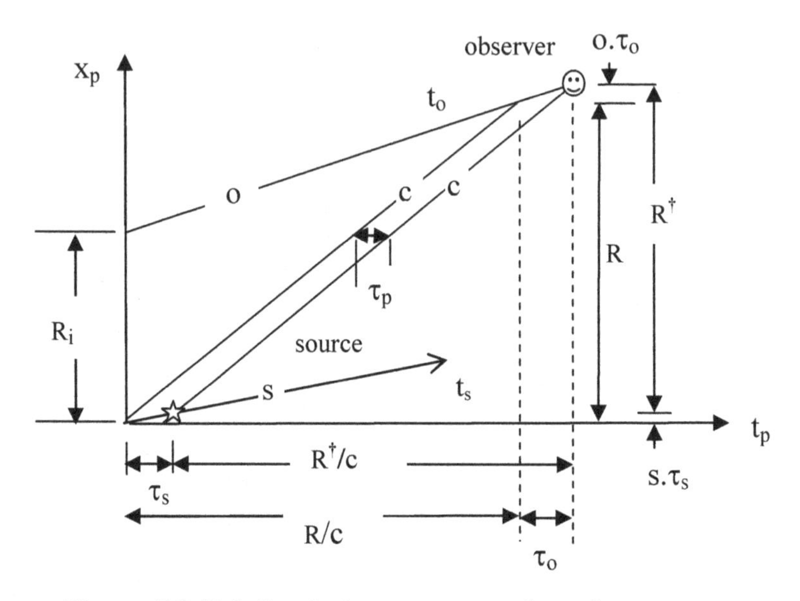

Figure 5.2 Relation between τ_s, τ_p and τ_o for source motion (velocity 's') and observer motion (velocity 'o') relative to a stationary propagation medium of time τ_p

$$t_p = \tau_s + R^{\dagger}/c = \tau_o + R/c \qquad (5.15)$$

and

$$x_p = s.\tau_s + R^{\dagger} = o.\tau_o + R \qquad (5.16)$$

Combining equations 5.15 and 5.16, eliminating $R^{\dagger}$ gives the required result, confirming equation 5.14, i.e.

$$(1\text{-}s/c)\,\tau_s = (1\text{-}o/c)\,\tau_o$$

Or in shorthand:

$$\tau_p = \varepsilon_s \tau_s = \varepsilon_o \tau_o, \quad \text{or} \quad K = \tau_o/\tau_s = \varepsilon_s \varepsilon_o^{-1} \qquad (5.17)$$

This event time ratio expresses the effect of both source and observer motion with respect to the propagation medium, based on the Galilean Transform. It establishes the purely classical event times of wave generation, transmission and reception in a propagation medium. K is the ratio of these source and observer times.

3 Relativistic (Lorentzian) Time

The classical wave propagation process has been considered in Section 2. What is required now is to include the effect of Lorentz's time and space contraction through motion, relative to the medium. It should be emphasized that it's not the stationary medium that changes through system motion. It's the electronic structure of matter, in the moving frame, that compresses, in space and time i.e. any kind of system, made of atoms or molecules, including life itself. The Lorentz two frame system transform (LT), is extended to three individual frames. This describes a full radiation model of both a source and observer moving around the universe, relative to the medium at rest in absolute space. The transform is illustrated in Figure 5.3.

The source and observer speeds with respect to the medium are considered again to be s and o, respectively. *The fundamental effect of Lorentz's system transform, is to maintain the invariance of the propagating speed* in the moving observer frame $c_o=x_o/\tau_o$ and the stationary propagation medium $c_p=x_p/\tau_p$. The theory is extended to include the moving source frame $c_s=x_s/\tau_s$. To accomplish the invariance of the speed of light, systems have to contract both in time and space in the moving frame, relative to the stationary propagation medium, by the same ratio α, exactly, i.e.

$$\tau_s = \alpha_s \tau_p, \quad \tau_o = \alpha_o \tau_p, \quad x_s = \alpha_s x_p, \quad x_o = \alpha_o x_p \qquad (5.18)$$

then

$$c_s = x_s/\tau_s = \alpha_s x_p/\alpha_s\tau_p = x_p/\tau_p = c_p \qquad (5.19)$$

$$c_o = x_o/\tau_o = \alpha_s x_p/\alpha_s\tau_p = x_p/\tau_p = c_p \qquad (5.20)$$

The relativistic time and space motional ratios from equation 5.18 are:

$$\tau_s/\alpha_s = \tau_p = \tau_o/\alpha_o \quad \text{or} \quad K = \tau_o/\tau_s = \alpha_o\,\alpha_s^{-1} \qquad (5.21)$$

and

$$x_s/\alpha_s = x_p = x_o/\alpha_o \quad \text{or} \quad K = x_o/x_s = \alpha_o\,\alpha_s^{-1} \qquad (5.22)$$

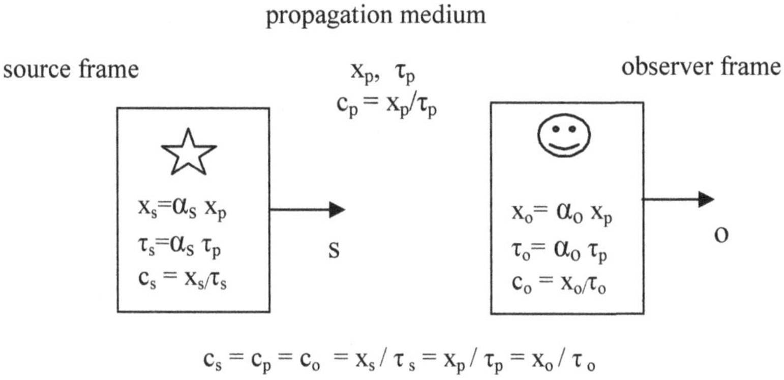

propagation medium

source frame

$x_p,\ \tau_p$

$c_p = x_p/\tau_p$

observer frame

$x_s = \alpha_s x_p$

$\tau_s = \alpha_s \tau_p$

$c_s = x_s/\tau_s$

s

$x_o = \alpha_o x_p$

$\tau_o = \alpha_o \tau_p$

$c_o = x_o/\tau_o$

o

$$c_s = c_p = c_o = x_s/\tau_s = x_p/\tau_p = x_o/\tau_o$$

Figure 5.3 Relativistic source and observer motion relative to a propagating medium at rest in space

Equations 5.21 and 5.22 give the time and space contraction of sources and observers moving with respect to the propagation medium. The stationary medium does not shrink with motion through it, i.e. x_p and t_p are unchanged. However, system time and space shrink in the moving frame, space in the direction of motion. The conventional Mach numbers for source (M_s) and observer (M_o) motion relative to the propagating medium are then:

$$s = (x_p/\tau_p)_s,\ \ M_s = s/c = (x_p/\tau_p)_s/c$$
$$(5.23)$$
$$o = (x_p/\tau_p)_o,\ \ M_o = o/c = (x_p/\tau_p)_o/c$$

Also hybrid (across frame) Mach numbers (M_s*) and (M_o*) between the stationary medium space x_p, and times τ_s and τ_o in the moving source and observer frames can be defined thus:

$$M_s{*}= (x_p/\tau_s)_s/c = (x_p/\tau_p)_s/c\alpha_s = M_s/\alpha_s,$$
$$(5.24)$$
$$M_o{*}= (x_p/\tau_o)_o/c = (x_p/\tau_p)_o/c\alpha_o = M_o/\alpha_o$$

Although it is presumed that the speed of light cannot be exceeded ($M<1$), within the medium frame, it appears that it can be exceeded across frames ($M^*>1$). The speed and distance capability is much greater in the hybrid frame by α^{-1}, where distance is measured in the stationary frame and the slower time in the moving frame. As $M\rightarrow 1$, $\alpha\rightarrow 0$ and $M^*\rightarrow\infty$, allowing huge distances to be achieved in space travel, in a short time in the moving frame, without exceeding the speed of light. See Section 4.3, Chapter VI for a worked example. 'Super-lightic' speed across frames $M^*=1$, occurs when $M\approx 0.7$. Space also contracts in the direction of motion, therefore astronauts and the space ship would need to rotate slowly to avoid permanent flattening in one direction.

4 Classical and Relativistic Time

Combining the classical and relativistic motional effects, assuming the effects are multiplicative, in accordance with the LT, one has from equations 5.17, 5.21, 5.22 the form of the motional event time ratio K:

$$(\varepsilon_s\tau_s)/\alpha_s=\tau_p=(\varepsilon_o\tau_o)/\alpha_o, \quad \text{or}$$
$$(5.25)$$
$$\tau_o/\tau_s = K = K_c.\ K_r = (\varepsilon_o^{-1}\varepsilon_s)_c\,(\alpha_o\,\alpha_s^{-1})_r$$

Equation 5.25 predicts the fractional event time change for both source and observer motion with respect to the medium. The classical and relativistic time change contributions are indistinguishable in motion. However, the former ceases (is temporary) the latter remains (is permanent) after motion stops. We can now establish the contraction fraction α. According to the Optical Principle of Relativity (OPR), there is no difference between an optical (electromagnetic) source moving towards a stationary observer and an observer moving towards a stationary source, assuming motion with respect to the propagating medium. To satisfy this principle, from equation 5.25, one has:

$$\tau_p/\tau_s = \varepsilon_s/\alpha_s\,, \quad \tau_o/\tau_p = \alpha_o/\varepsilon_o\,, \quad \tau_p/\tau_s = \tau_o/\tau_p \qquad (5.26)$$

$$\text{giving} \quad \varepsilon_s/\alpha_s = \alpha_o/\varepsilon_o$$

where

$$\varepsilon_s = 1\text{-}M_s\cos\sigma_s \ \text{ and } \ \varepsilon_o = 1+M_o\cos\sigma_o$$

Assuming OPR is valid only for direct source and observer motion ($\sigma_s = \sigma_o = 0$) and removing the distinction between source and observer motion for equal speeds i.e. $M_s=M_o=M$, then the relativistic contraction from equation 5.26 becomes:

$$\alpha = (\varepsilon_s\,\varepsilon_o)^{1/2} = [(1\text{-}M)\,(1+M)]^{1/2} = (1\text{-}M^2)^{1/2} \qquad (5.27)$$

Equation 5.27 is the general time and space Lorentz contraction. It is also called the Fitzgerald (1889) spatial contraction, which incidentally for space was obtained 10 years before Lorentz's general result. For non-direct approaches, flybys and other flight manoeuvres, $\sigma \neq 0$, and as α is omni-directional and ε directional, then the total effect will no longer be symmetrical i.e. an approaching source event time will no longer be equal to an approaching observer time for the same speed, i.e. source and observer motions are distinguishable, apart from a direct approach. It's possible for relative speeds between systems to exceed light speed, with respect to the medium, but not directly between each other, determined by relativistic addition of velocities.

Away from gravitational matter the EM motional (wave equation) analysis (EMMA) gives the observer time τ_o in terms of the source event time τ_s. From equations 5.25 the generalized classical transform for both source and observer motion, modified by the Lorentz contraction, gives the relationship between these event times for a given source and observer flight path as:

$$\tau_o = \int [\varepsilon_s \alpha_s^{-1} \varepsilon_o^{-1} \alpha_o]_{tp} \, d\tau_s \qquad (5.28)$$

Where, t_p outside the square bracket emphasises that all motions are evaluated in the stationary propagation medium time and space. At low speeds, $\alpha_s \approx \alpha_o \approx 1$, $\alpha_s^{-1}\alpha_o \approx 1$ giving:

$$\varepsilon_s\varepsilon_o^{-1} = (1\text{-}M_s\cos\sigma_s)/(1+M_o\cos\sigma_o) \approx 1\text{-}\ (M_s\cos\sigma_s\text{-}M_o\cos\sigma_o)$$

For inline approaching systems, $\sigma_s=\sigma_o$,

$$\varepsilon_s\varepsilon_o^{-1} \approx 1\text{-}(M_s\text{-}M_o) \qquad (5.29)$$

The effect reduces to a relative velocity between the source and observer. A result relativists would attempt to argue is valid for all speeds. Neglecting the temporary Doppler time changes i.e. $\varepsilon_s=\varepsilon_o=1$, the accumulative permanent time changes for incremental changes and $M<<1$, from equation 5.25 and 5.27 one has:

$$\tau_s=\alpha_s\alpha_o^{-1}\tau_o,\ \ \Delta\tau_s=(\Delta\alpha_s+\Delta\alpha_o^{-1})\Delta\tau_o,\ \ \Delta\alpha_s=\text{-}M_s^2/2,\ \ \Delta\alpha_0=M_o^2/2 \qquad (5.30)$$

i.e. the source time contracts (slows) (negative) through source motion and expands (quickens) (positive) through observer motion, compared with the propagation time.

When the source and observer are not in contact initially, i.e. there is an initial separation distance, then there will be an initial propagation time (retarded) delay before the observations are seen. From equation 5.13 and Figure 5.1, the observer time then becomes:

$$t_o = R_o/c+\tau_o, \quad R_o=\alpha_o R_p \qquad (5.31)$$

R_p is the source-observer initial propagation distance in the medium and the R_o is propagation distance in the moving frame, which contracts through motion by $\alpha_o R_p$. For source and observer motion not in the same direction, the directional angles σ_s and σ_o are made with the directions of motion and a straight line joining the source emission and observer reception positions, as illustrated in Figure 5.1. For different source and observer speeds, in different directions, equation 5.28 gives the result. Observationally, there is nothing extraordinary regarding arbitrary motion. In this medium based theory, time changes instantly at the source emission and observer reception times according to their speeds at that instant in the medium, whether it is constant or accelerative. This is in contrast to Einstein's inertial frame, which implies there is no difference between a stationary and constantly moving system.

5 New Space -Time

Equation 5.28 gives the relation between the source event time τ_s, and the observer event time (observed rate of happening) τ_o, for arbitrary source and observer flight paths through the universe. Its validity and authority is based on absolute source and observer motion with respect to the propagating medium. The directional classical time change, affected by $\varepsilon_s\varepsilon_o^{-1}$, is caused through the PTA wave dynamics, including the generation, transmission and reception of the EM waves. The omni-directional Lorentzian time change, determined by $\alpha_s^{-1}\alpha_o$, where $\alpha_s=(1-M_s^2)^{1/2}$ and $\alpha_o=(1-M_o^2)^{1/2}$, is caused through the relativistic contraction (both time and space) in the moving frame, through motion relative to the propagation medium. For M<<1, $\alpha_o \approx 1-M_o^2/2$ or $\alpha_o^{-1} \approx 1+M_o^2/2$. In the case of discrete source and observer event time segments τ_{s1} τ_{s2} τ_{s3} etc. and τ_{o1} τ_{o2} τ_{o3} etc., respectively, equations 5.28 and 5.31 become:

$$t_o = R_o/c + \tau_o = \alpha_o R_p/c + \left[\frac{\varepsilon_s}{\alpha_s}\frac{\alpha_o}{\varepsilon_o}\Delta\tau_s\right]_{\tau s1,\tau o1}^{\tau s2,\tau o2} + \left[\frac{\varepsilon_s}{\alpha_s}\frac{\alpha_o}{\varepsilon_o}\Delta\tau_s\right]_{\tau s2,\tau o2}^{\tau s3,\tau o3}$$

$$+ \left[\frac{\varepsilon_s}{\alpha_s}\frac{\alpha_o}{\varepsilon_o}\Delta\tau_s\right]_{\tau s3,\tau o3}^{\tau s4,\tau o4} + \left[\frac{\varepsilon_s}{\alpha_s}\frac{\alpha_o}{\varepsilon_o}\Delta\tau_s\right]_{\tau s4,\tau o4}^{\tau s5,\tau o5} \quad \text{etc} \qquad (5.32)$$

The source and observer flight paths and the three event time scales τ_s, τ_p and τ_o are illustrated in Figure 5.4. The vertical and horizontal axes represent absolute time t_p and space x_p, in the medium respectively. Corresponding event times, in the source emission time τ_s and observer reception time τ_o, are connected by propagation light paths. These paths represent light moving forward in time, from source to observer, going from left to right in the direction of the arrows, whose initial propagation delay in the medium, from equation 5.31, is $\alpha_o R_p/c$ and whose propagation gradient is 1/c (45° if x_p is in light years). If the source and observer were interchanged, the light paths would go from right to left and have a negative gradient, giving a totally different observed time history for the same flight paths. Ether-less relativity cannot distinguish between a constantly moving and stationary system, or between source and observer motion, whereas Figure 5.4 does distinguish between these situations.

Equations 5.28, and 5.31 and their illustration in Figure 5.4, are the results of this chapter. In the absence of gravity, General Relativity has no part. The space-time flight paths give the relation between the various time scales through source and observer motion relative to the propagation medium. The motional history of the source and observer is therefore indelibly imprinted in the observed time history. The gradient of the source and observer flight paths cannot exceed ±1/c. A zero gradient represents infinite speed and conversely an infinite gradient indicates zero velocity. Actually, the classical space–time diagram should be tilted on its side so that time is horizontal and then all gradients would be speeds not inverse speeds. In these particular flight paths, the τ_s and τ_o curves start at the same time but at a different place. They end at a different time, but at the same place. Examples using equation 5.28 and Figure 5.4 are given in Chapter VI.

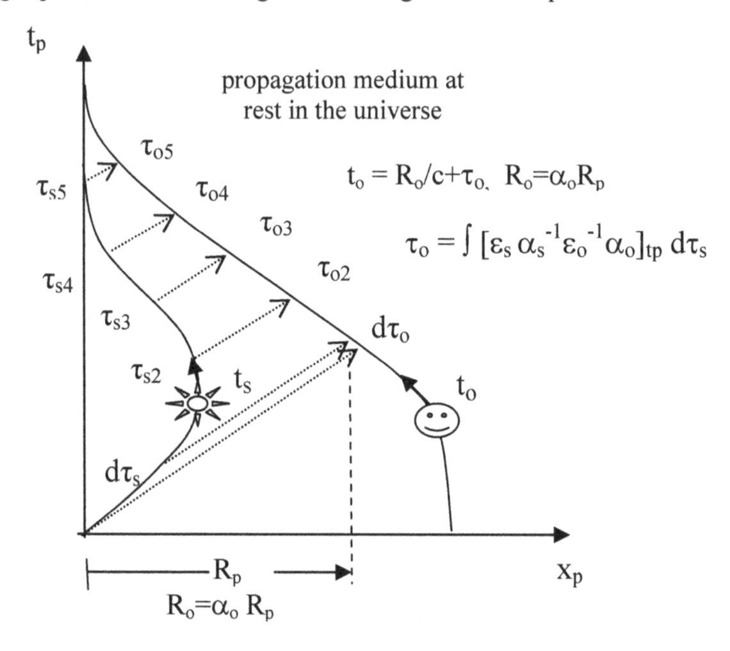

Figure 5.4 Space-time diagram, showing the relation between source time t_s and observer time t_o relative to the medium time t_p, showing summation of the time segments for source and observer flight paths

6 Gravity and Frame Motion

A link is now made between the Lorentz transform, accelerating frames and gravity. All three result in permanent time and space changes. *They are connected with motion relative to, and compression of the propagation medium.* According to the Lorentz transform space contracts in the direction of motion and time slows moving relative to the medium. This leads to the source contracting (time slowing) through motion. Whereas, observer contracting leads relatively to the surrounding medium expanding (source time quickening). Basically, there is nothing distinctive regarding system acceleration; time slowing is concerned with the system's actual velocity at any instant with respect to the medium.

Individual source and observer motion has been considered for both constant and arbitrary motion with respect to the stationary propagation medium. The observed event time for both motions is given by equation 5.28, and illustrated in Figure 5.5(a).

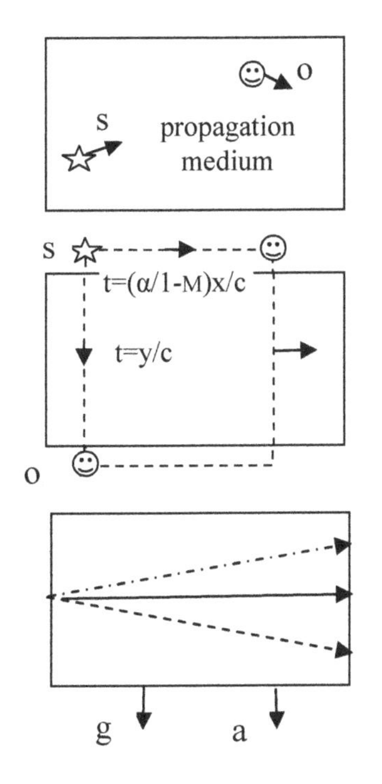

(a) Individual source and observer motion relative to the medium, source motion contracts, observer motion expands event time and space.

(b) Source and observer motion (fixed separation) contract systems in the direction of motion, making propagation time variant, but invariant perpendicular to motion.

(c) Gravity compresses medium, observer acceleration effectively expands medium. Ray deflects down towards gravity 'g', and up through acceleration 'a'. Ray is horizontal when a=g (freely falling).

Figure 5.5 (a) Individual source and observer motion. (b) Fixed separation source and observer motion and (c) Effect of observer acceleration 'a' and gravity 'g' on a horizontal ray

Consider now a system consisting of a source and observer, of fixed separation, moving together relative to the propagation medium. They are moving in a straight line, as shown in Figure 5.5(b). In Einstein's Inertial Frame (IEF), the propagation time perpendicular to and in the direction of the frame motion is invariant, as discussed in Section 3.2, Chapter IV. However, unlike EIF, time in the direction of motion, in the revised theory, is variant through motion by $\alpha/(1\text{-}M)$, as discussed in Section 3.3, Chapter IV. Thus, propagation time is invariant normal to the motion but variant in the direction of motion, according to the LT. Also there is no event time change between the source and observer through constant motion at any speed, i.e. $\varepsilon_s\alpha_s^{-1}\varepsilon_o^{-1}\alpha_o=1$, in equation 5.28.

Consider further an actual light beam propagating in the medium in the stationary frame. In the presence of a gravitational field 'g', the medium compresses, according to Schwarzschild (1916). Space contracts and time slows, bending a horizontal light ray in the direction of the field g, downwards, as depicted in Figure 5.5(c). If now the observer frame is accelerated downwards by 'a', in the direction of the field, time slows and space contracts at the observer. This makes the surrounding medium appear to expand, and its time quicken, relatively, making a light ray appear to bend upwards (away from the direction of motion), in the moving frame.

Thus, gravity compresses the medium (time and space contract) in the stationary medium in the direction of the field, and observer motion (equation 5.30) effectively expands the medium (time and space effectively expand in the observer frame). When the frame acceleration (expansion), as in a falling system (lift), is equal to the gravitational field (compression) in the stationary medium a=g, (freely falling), the ray will again appear horizontal. This is an optical or time view of the Equivalence Principle, based on a moving observer and a stationary medium. It depends on a clear distinction between source and observer motion, which ether-less SR does not make.

It's obvious in a lift freely falling that everything in it will also freely fall giving no relative motion (all objects are attracted and accelerate together towards the Earth). But for objects to be observed as freely falling (floating) and times to be neutral, the compression of the medium through gravity must be equalled by the relativistic expansion through observer motion. It is not that gravity attracts light; it is the medium through which the light travels that is compressed. If there was no medium there could be no gravitational compression, or observer acceleration expansion.

7 Gravitational and Accelerative Time

7.1 Gravitational time

The source event time $\tau_{g,}$ in a gravitational field, compared to the general free field propagation value τ_p, can be found in a manner similar to the previous time scales format. Here, the medium and systems close to a gravitational mass compress (space contracts, time slows), according to Schwarzschild (1916), the dominant term is:

$$\tau_{g\_} = (1\text{-}2Gm^*/c^2R_g)^{1/2}\,\tau_p \qquad (5.33)$$

Using a gravitational operator δ_g, for M<<1:

$$\tau_g = \delta_g\tau_p,\ x_g = \delta_g x_p,\ \delta_g=(1\text{-}M_g)^{1/2}\approx(1\text{-}M_g/2),\ \Delta\tau_g=\Delta\delta_g\Delta\tau_p=-(M_g/2)\,\Delta\tau_p \qquad (5.34)$$

$$M_g=2Gm^*/c^2R_g \text{ or } 2gR_g/c^2 \text{ as } g=Gm^*/R_g^{\ 2}$$

M_g is a gravitational Mach number (effective velocity over the speed of light), g is the acceleration due to gravity (at the surface of the Earth g=9.8 m/s^2). G is the gravitational constant 6.67×10^{-11}Nm^2/kg^2 (G/c^2= $7.4\text{x}10^{-28}$), m* is the gravitational mass of the body in kg and R_g is the radial distance from the body centre. Unlike Newton's gravitational (field intensity) inverse square law with distance, (last term in equation 5.34), Schwarzschild medium compression (time ageing) only reduces according to the inverse law, similar to potential. The event horizon (radius) from a gravitational matter singularity (black hole), inside of which light cannot escape, is then given when M_g=1 (gravitational speed is equal to the light speed). From equation 5.34, one has:

$$R_g=2Gm^*/c^2 \qquad (5.35)$$

From equation 5.34, reduced time ageing (time slowing) through gravitational medium compression, is given approximately by $\Delta\tau_g = \Delta\delta_g\tau_p$ where $\Delta\delta_g \approx -M_g/2=-Gm^*/c^2R_g$. For example on the Earth's surface, $m^*=5.98\text{x}10^{24}$kg, $R_g=6.37\text{x}10^6$m, $\Delta\delta_g=-(G/c^2)m^*/R_g=-7.4\text{x}10^{-28}\text{x}$ $5.98 \text{ x } 10^{24}/6.37 \text{ x}10^6 =-7\text{x}10^{-10}$ and for a 24 hour period, $\tau_p=8.6\text{x}10^4$s, $\Delta\tau_g=-7\text{x}10^{-10}\text{x}8.6\text{x}10=-6\text{x}10^{-5}$s, i.e. 60 μs/day slowing on the Earth's surface, compared to free space. Also at the Earth's distance of 1.5x 10^{11}m, from the Sun of mass $1.99\text{x}10^{30}$ kg, $\Delta\delta_g=-(G/c^2)m^*/R_g=-7.4\text{x}$ $10^{-28} \text{ x}1.99\text{x}10^{30}/1.5\text{x}10^{11}=-9.8\text{x}10^{-9}$ and for a 24 hour period, $\Delta\tau_g=-9.8\text{x}10^{-9}\text{x}8.6 \text{ x}10^4=-7.4\text{x}10^{-4}$. Thus the time slowing on the Earth's surface due to the Sun's presence is 740 μs/day, i.e.≈X10 slower than the Earth's gravitational time slowing of 60 μs/day.

A gravitational fractional time change across a small distance $\Delta R = R_1-R_2$, in a decreasing field, $R_1>R_2$, from equation 5.34 becomes:

$$\Delta\tau_g/\tau_p=\Delta M_g/2=(Gm/c^2)(1/R_2-1/R_1)\approx(Gm/c^2)(R_1-R_2)/R^2=g\Delta R/c^2 \qquad (5.36)$$

For example, ΔR=10km vertically above the Earth $\Delta M_g/2=g\Delta R/c^2= 9.8\text{x}10^4/(3\text{x}10^8)^2=1.1\text{x}10^{-12}$, and for a 24 hour period, $\Delta\tau_g=-1.1\text{x}10^{-12}\text{x}$ $8.6\text{x}10^4=-95\text{x}10^{-9}$s, i.e. a difference of 95ns/day quickening. Also light propagating a distance ΔR against or with a gravitational field is stretched or compressed, its wavelength changing by $\Delta\lambda$, giving red shifted or blue shifted light respectively. From equation 5.36:

$$\Delta\tau_g/\tau_p=\Delta\lambda/\lambda=\Delta M_g/2=g\Delta R/c^2 \quad (5.37)$$

7.2 Observer accelerative time

Now from equation 5.21, $\tau_s=\alpha_s\tau_p$, $\tau_p=\alpha_o^{-1}\tau_o$, i.e. a moving source time τ_s compared to the propagation medium time τ_p contracts by α_s, whereas a moving observer time τ_o compared to the propagation medium time τ_p expands by α_o^{-1}. This means that the time slowing through gravity can be effectively reduced through observer motion. If an observer of acceleration 'a' moves a distance d, the time t for a light wave to reach d is d/c. The velocity attained by the observer in this time is v=at=ad/c. The Mach number is then $M=v/c=ad/c^2$. For M<<1, the effective time change in the medium through observer acceleration, $\Delta(\tau_p)_o$, in the format of equation 5.34, is from equation 5.30, for a stationary source $\alpha_s=1$, $\tau_s=\tau_p$, is then:

$$\Delta(\tau_p)_o = \Delta\delta_o\,\Delta\tau_o, \quad \Delta\delta_o = M_o, \quad \text{where } M_o = ad/c^2 \quad (5.38)$$

7.3 Total time

Equation 5.38 is equivalent to equation 5.36, where $M_o=M_g/2$, a=g, and d=ΔR. Thus the effect of medium compression through gravity and expansion through observer acceleration, are equivalent. Acceleration and gravity are dependant on M, whereas constant motion (LT), from equation 5.30, is dependent on M^2. Thus all three effects, gravity, accelerating frames and the LT are related through an equivalent velocity relative to the propagation medium. One can now obtain, in a simple potential addition, the total time change $\Delta\tau_T$, caused through constant source motion $\Delta\tau_s$, or through acceleration, either, observer $\Delta(\tau_p)_o$ or gravity $\Delta\tau_g$ as:

$$\Delta\tau_T = \Delta\tau_s+\Delta(\tau_p)_o+\Delta\tau_g \quad (5.39)$$

Where for M<<1:

Constant source motion $\Delta\tau_s = \Delta\alpha_s \Delta\tau_p$, $\Delta\alpha_s \approx -M_s^2/2$, $M_s = s/c$, (5.40)

Observer acceleration $\Delta(\tau_p)_o = \Delta\delta_o \Delta\tau_p$, $\Delta\delta_o \approx M_o$, $M_o = ad/c^2$ (5.41)

Gravity $\Delta\tau_g = \Delta\delta_g \Delta\tau_p$, $\Delta\delta_g \approx -M_g/2$, $M_g = 2Gm^*/c^2R_g = 2gR_g/c^2$ (5.42)

This time equivalence equation (5.39) predicts time slowing (-) through constant source motion (LT) (first term), time quickening (+) through observer acceleration (second term), and time slowing (-) at a fixed point in a gravitational field (third term). The time Equivalence Principle is given by the last two terms. For free fall, $M_o=M_g/2$, gravity is neutralized.

Generally, from symmetry, for both source and observer motion in different gravitational fields, the relation between the observer time τ_o and source time τ_s, in terms of classical, relativistic and gravitational effects, the form of the total event time ratio (K_T) is:

$$(\varepsilon_s/\alpha_s\delta_s)\tau_s = \tau_p = (\varepsilon_o/\alpha_o\delta_o)\tau_o \quad (5.43)$$

$$\text{or} \quad \tau_o/\tau_s = K_T = K_c K_r K_g = (\varepsilon_o^{-1}\varepsilon_s)_c(\alpha_o \alpha_s^{-1})_r(\delta_o \delta_s^{-1})_g \quad (5.44)$$

Where δ_o, δ_s are the observer and source gravitational transforms in their respective gravitational fields. Equation 5.44 predicts the total event time transform for the additional effect of source and observer motion of different constant speeds in different gravitational fields.

8 Conclusions

The EM Motional Analysis (EMMA) has been derived for EM sources and observers moving relative to the EM medium. The analysis is developed directly from LT (1899). It also includes the effect of gravity using Schwarzschild metric (1916), both effects can be considered to be independent of SR (1905) and GR (1915). The resulting analysis allows individual source and observer flight paths to be chartered on a common universal propagation medium continuum.

The new theory establishes an event time transform between the source time τ_s and that what is observed τ_o at the observer, for arbitrary source and observer flights through the universe. Systems change (time and space contract) through motion relative to the medium, distinguishing clearly between stationary and constantly moving systems. Interchanging the source and observer for the same flight paths gives different observations. Einstein's ether-less SR cannot distinguish between these situations. The distance capability across the universe is much greater using hybrid reference frames. Here light speeds can effectively be exceeded without exceeding the speed of light in the propagation medium.

Gravity compresses both the medium and sources within it. An accelerating observer relatively expands the medium and sources within. Thus gravity can be increased, reduced or neutralised, through observer acceleration, explaining optically The time Equivalence Principle. Gravity does not attract light; it is the medium through which the light travels that is compressed. If there was no medium there could be no gravitational compression or observer acceleration expansion. Gravity, observer acceleration, and the LT are all related through a equivalent velocity relative to the propagation medium.

Chapter VI:

Applying the Theory

1 Introduction

In this Chapter the detailed properties of the event time transform are investigated i.e. how the source event time appears to an observer, as the source and observer move relative to the medium. It includes the effect of relativistic and classical motion. The transform determines whether there is absolute time and space, whether it is possible to time travel, or whether there is time slowing through source motion or time quickening through observer motion. Also the event time delay between the source emission and the observer reception positions is important. Here the difference between broadside and in line motion is considered. A new space-time is established where source and observer flight path observations can be compared on the same Minkowsky diagram. Both event time and propagation time changes are important in investigating EM sources and observers in motion. Practical examples of sources and observers in motion are analysed.

2 Event Time Transform

For motion, away from large masses, the observer event time τ_o in terms of the source event time τ_s, is determined by EM Motional Analysis (EMMA), derived in Chapter V, equations 5.28 and 5.31, The total integrated or accumulative observer time scale in terms of the source time is given below and illustrated in Figure 5.4.

$$t_o = \tau_o + R_o/c, \qquad R_o = \alpha_o R_p$$
$$\tau_o = \int [\varepsilon_s \alpha_s^{-1} \varepsilon_o^{-1} \alpha_o]_{tp}\, d\tau_s \qquad (6.1)$$

Where t_p outside the square bracket indicates that quantities inside the bracket are to be evaluated in the propagation medium time t_p. t_o is the time in the moving observer frame. R_o and R_p are the initial propagation distances in the moving and stationary frames, respectively. From equation 6.1 the instantaneous motional event time transform becomes:

$$K_t = (\tau_o/\tau_s) = \varepsilon_s \alpha_s^{-1} \varepsilon_o^{-1} \alpha_o = K_{\varepsilon s,\alpha s} K_{\varepsilon o,\alpha o} \qquad (6.2)$$

The individual source and observer event time transforms, at their respective emission and reception times and positions in the medium, are $K_{\varepsilon s,\alpha s} = \varepsilon_s \alpha_s^{-1}$, and $K_{\varepsilon o,\alpha o} = \varepsilon_o^{-1} \alpha_o$. The relativistic operators are $\alpha_s = (1-M_s^2)^{1/2}$, $\alpha_o = (1-M_o^2)^{1/2}$, and the classical Doppler operators $\varepsilon_s = 1-M_s \cos \sigma_s$, $\varepsilon_o = 1-M_o \cos \sigma_o$, where $M_s = s/c$ and $M_o = o/c$. The directional angles σ_s and σ_o are made with the directions of motion and a straight line joining the source emission and observer reception positions, as illustrated in Figure 5.1. The total observer event time τ_o, at the end of any given flight path, in terms of the source event time τ_s, depends on the initial propagation distance R_o and on how K_t (the individual motional operators ε_o^{-1}, α_o, ε_s, and α_s^{-1}) changes with time during the source and observer journeys through space.

As a simple well known example, consider a moving source and stationary observer, $M_o=0$, $\varepsilon_o^{-1}\alpha_o=1$. The total integrated observer time around the source flight path, from equation 6.1, then becomes $\tau_o = \int [\varepsilon_s \alpha_s^{-1}]_{tp}\, d\tau_s$. For a closed flight path, the directional classical operator ε_s integrates to unity, it is also unity if the motion stops. This leaves only the omni-directional relativistic operator α_s^{-1} accumulative time changes, which lead to the famous twin time paradox. However, there is no ambiguity here in deciding which twin becomes younger. The twin who actually journeys around the universe, relative to the propagating medium, becomes younger than the twin who stays at home (stationary with respect to the medium).

Figure 6.1 shows the instantaneous motional event time transform, $K_t = K_{\varepsilon s,\alpha s} = \varepsilon_s \alpha_s^{-1}$, for source motion only (full curve), with a stationary observer i.e. $K_{\varepsilon o,\alpha o} = \varepsilon_o^{-1} \alpha_o = 1$. The individual classical $K_{\varepsilon s}$ and relativistic $K_{\alpha s}$ components are also shown (dotted straight line and chain dotted symmetrical curve, respectively). For K_t values greater than unity $\alpha_s < 1$, $\alpha_s^{-1} > 1$, $(\tau_o > \tau_s)$, the source time τ_s contracts, going at a slower rate than the observer time τ_o. This is referred to as red shifted, referring to the observed lowering of the source frequency. Whereas, for $K_t < 1$, $(\tau_o < \tau_s)$, the source time τ_s expands, quickens compared to the observer time τ_o, known as blue shifted. The propagation time history for this example is considered further in Section 4.3.

It can be seen in Figure 6.1, at ± low speeds, that the relativistic effect is negligible and the classical effect (straight line) dominates. At high speed, $M \rightarrow 1$, $K_{\alpha s} \rightarrow \infty$ and the source time slows to a stop (stands still). On the receding source side (M negative), the relativistic effect $K_{\alpha s}$ reinforces and dominates the classical effect $K_{\varepsilon s}$, making the observed time $(K_{\varepsilon s,\alpha s})$ red shifted, more than just the Doppler effect. However, on the approaching source side (M positive), although there are infinities and zeroes involved, the classical effect $K_{\varepsilon s} \rightarrow 0$ apposes and dominates the relativistic effect $K_{\alpha s} \rightarrow \infty$ at all speeds, including $M=1$, making the observed time $(K_{\varepsilon s,\alpha s} \rightarrow 0)$ blue shifted (speeded up).

At the observer there is no way to distinguish between instantaneous classical and relativistic time contributions; both are equally valid event time changes. Consider the proverbial travelling twin returning from visiting a distant space station. Although the relativistic effect (red shifted), on approaching home is large, the net effect for both time changes is an overwhelming classical time domination (blue shifted). Thus against popular belief, the approaching twin's time will appear through a telescope to be blue shifted (speeded up, increased ageing). But through the communication video, providing the communication frequency is automatically aligned, the twin will be seen as ageing less.

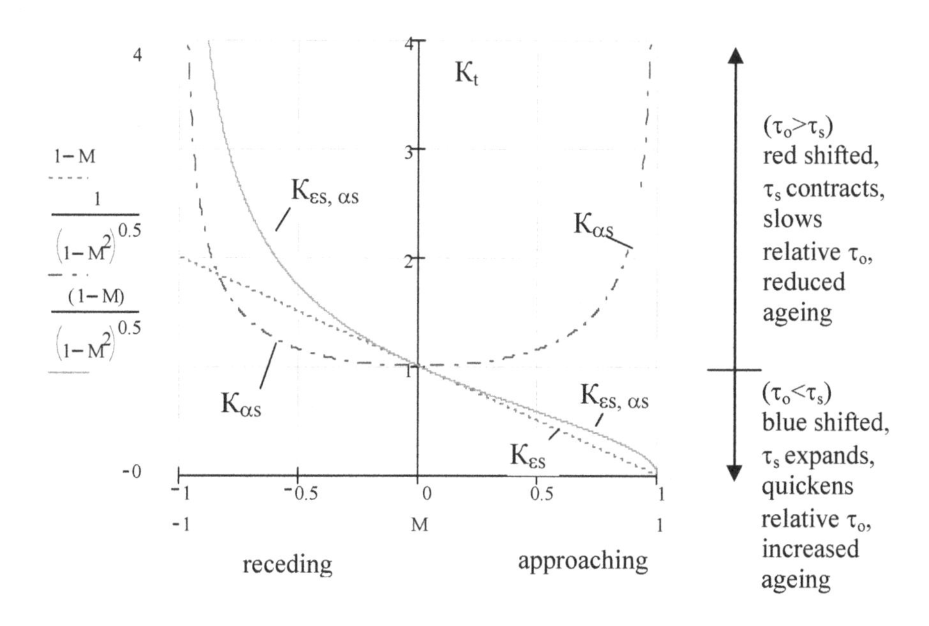

Figure 6.1 Instantaneous motional event time transforms K_t for EM source motion relative to a stationary observer and medium

Of course, when the travelling twin arrives, and his motion stops, the classical blue shifting stops, leaving an accumulative relativistic age slowing over the entire journey, making the travelling twin actually younger than his stay at home twin. The opposite is true, of course, for a receding source shown on the left hand side of Figure 6.1, where both the classical and relativistic source time slows down. The problem dramatically increases as the speed of light is approached.

3 Direction of Time

Considerable speculation has arisen through time travel in recent years. One should first ask what is meant by time? Time here is considered to be a means of measuring and comparing the order and speed of events. The colloquial expression says that time goes marching on, clock times stationary in the medium certainly do that, they move forward at an equal and constant rate. They also slow moving through the propagation medium and in gravitational fields. Further, the source event time has to be transmitted from the source to the observer, via the propagation medium, before it can be realised at the observer, in accordance with cause and effect. Thus, the observed source event time depends on source and observer motion with respect to the medium, which can slow down or speed up the observed source events. Time is therefore an observed effect. Without a medium, there is no physical mechanism, no wave propagation, no solution of the wave equation, no observations, no direction of time and no ordered cause and effect.

In this revised theory, material (substance transmission) time travel into the future or past is not possible, it is not in agreement with causality. To actually participate, change the source events as they happen, the observer must come in physical contact (interact) with the source, *at the source event time.* By definition this rules out any kind of distant participatory time travel. Causally, it is possible to observe past events, but not to participate in them, because they have already happened. It is not possible (non causal) to observe or participate in source events in the future, because they have not yet occurred.

Through motion or visiting a large mass, a traveller can slow his/her time down compared with a stationary system time and return to its future, actually participating in its present. This is quite feasible, but it is not reversible time travel. The process is one-way; one cannot reverse the flow of time in this situation. All that can be achieved is to alter the rate of ageing (time slowing) of one system compared to another. This also could be true within black holes and connecting worm holes, if they exist. Space will shrink to practically nothing and time will almost stop, allowing great distances to be achieved in little time. This assumes black holes can be exited at the other end of the journey to realise any benefit. However, this is still not time travel, it is just getting there quicker and ageing less.

It is possible, in principle, to visit the past, visually, and return, but not to interact or interfere with it. This would be possible if the speed of light could be exceeded (M>1). Unfortunately, at the speed of light, mass becomes infinite making this speed questionable. Therefore, any prediction for speeds greater than the speed of light must be highly speculative. If this critical speed could be avoided or bypassed, then the properties again become well behaved (finite and causal). Whether or not 'super-lightic' speeds are possible, one can predict their properties, according to this revised theory. To solve the wave equation, and classify the order of events, an EM motional event time transform K_t (equation 6.2), is used. It has positive values representing events going forward and negative values going backwards in time.

As a different example, consider now a moving observer with the source at rest. M is now positive for a receding observer. Observations are reversed compared to a moving source. A contracting moving observer relatively makes the source and surroundings expand. The EM motional event time transform $\tau_o/\tau_s=K_t=K_{\varepsilon o,\alpha o}=\alpha_o\varepsilon_o^{-1}$, from equation 6.2, is given by the solid curve in Figure 6.2. It shows a range of observer speeds, including greater than the speed of light, $|M|>1$. It can be seen that the relativistic time transform $K_{\alpha o}$ (chain dotted) is positive and symmetrical about M=0. It has singularities (zeros) at M=±1, indicating infinitely large (fast) observer time. For $|M|>1$, $K_{\alpha o}$ is again finite, showing observer time expanding (slowing), but it is now curiously imaginary, which is interpreted as an oblique time running at 90° to our own.

The relativistic time transform has no possibility of becoming negative i.e. enabling time to go backwards. On the other hand, the classical time transform $K_{\varepsilon o}$ is not symmetrical about M=0. It has an infinity for the receding observer at M=+1, indicating infinite small observer time (source time stopping). So for M<1 source time slows and for M>1 the time transform becomes negative reversing the source time, leaving the present and going back to the past. In the approaching observer case, M negative, there is no negation of the motional time transform and therefore no going back in time. Source time will speed up compared to observer time, becoming infinitely large (fast) at M=-1 and then becoming finite again for M<-1.

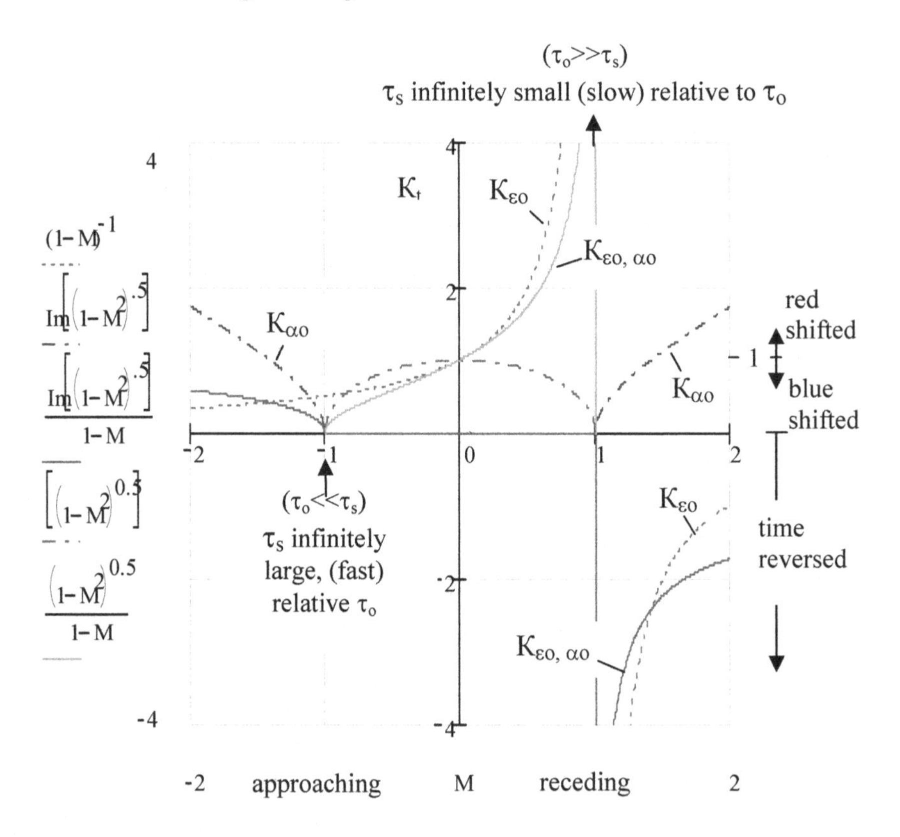

Figure 6.2 Instantaneous event time transforms K_t for 'super-lightic' observer speed $|M| > 1$ relative to a stationary source and medium

Thus for a receding observer, there is a potential to go back in time and observe the past (it is causal), but not to participate or interfere with it. One cannot go back in time and terminate his mother's life, when she was a child, as this would result in him or her not being born, which is clearly absurd. However, if the speed of light could be exceeded or bypassed, perhaps a traveller could go back to the past, see a broken cup reassemble itself, see an exploding supernova imploding, or more macabre, see ancestors resurrecting from their graves. Perhaps even for $M<1$, a cusp in a non-linear (curved) flight path could also give glimpses of the past.

In summary, in this revised theory, it is not possible (non causal) to travel materially into the future or past. It is possible (causally) to travel visually to the past, but not to interfere or change it. It is possible to slow one's time down through material transport, by physically moving at a high speed or visiting a massive body. However, again this is not reversible time travel, it is just relatively controlling the rate of ageing. Finally, unlike classical theory where moving source and observer Doppler factors are not the same, it should be noted from Figures 6.1 and 6.2, that through the Lorentz transform, the total observed event time for both source ($K_{\varepsilon s,\alpha s}$) and observer motion ($K_{\varepsilon o,\alpha o}$), is identical for inline motion, also confirmed through equation 2.3, in Section 7, Chapter II.

4 Propagation Time

The motional event time transform, $K_t=\varepsilon_s\,\alpha_s^{-1}\varepsilon_o^{-1}\alpha_o$, gives information regarding the observed event time changes in terms of the source event time. However, changes in propagation times R_o/c, between the source emission and observer reception position in the retarded time equation $\tau_o=t_o-R_o/c$, in equation 6.1, are equally important. Within frames, where both source and observer, at a fixed separation, move together at constant velocity in a straight line, there is no event time change,

$\varepsilon_s\alpha_s^{-1}\varepsilon_o^{-1}\alpha_o=1$. However, there is a propagation time change depending on the source and observer motion and orientation of the propagation path in the medium compared with the direction of motion. This propagation time variation, is an important tool in investigating the properties of sources and observers in motion.

4.1 Broadside propagation

Consider, for example, a diamond shaped formation of five equi-spaced space ships, labelled A to D, separated by distance 'b' from the centre mother ship. The ships (sources and observers) are moving together at the same speed 'o' and direction relative to the propagating medium, as illustrated in Figure 6.3. Consider first, broadside (sideways) propagation to the direction of motion, as in communication of say from centre ship B to adjacent ship E to the right. The situation is illustrated in the space-space diagram in Figure 6.4, where the dotted lines indicate the actual light paths in the propagation medium from source to observer.

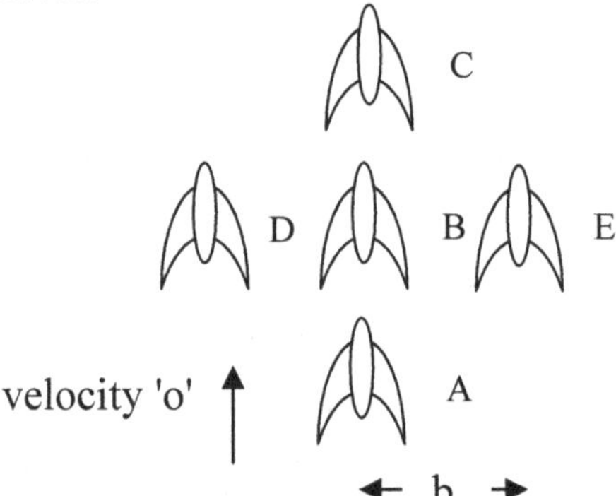

Figure 6.3 Space ship formation, propagation time in the direction of motion varies with speed, but is independent of motion to the side

φ is the propagation angle made between the direction of motion and the light propagation from the source to observer. 'a' is the distance travelled in the direction of motion. φ_p and a_p are the actual values in the propagation medium, and φ_o and a_o are the values in the moving frame. Classically the light has to propagate forward in the medium, with propagation angle φ_p, to be able to communicate with the ship at position E' and then back to position B''. Alternatively, from the ship's moving frame point of view, this motion could be considered as a head wind for forward propagation or a cross wind for broadside propagation. For the source and observer moving together the event time between the source and observer for any angle will not change i.e. $K=\varepsilon_s \alpha_s^{-1} \varepsilon_o^{-1} \alpha_o=1$.

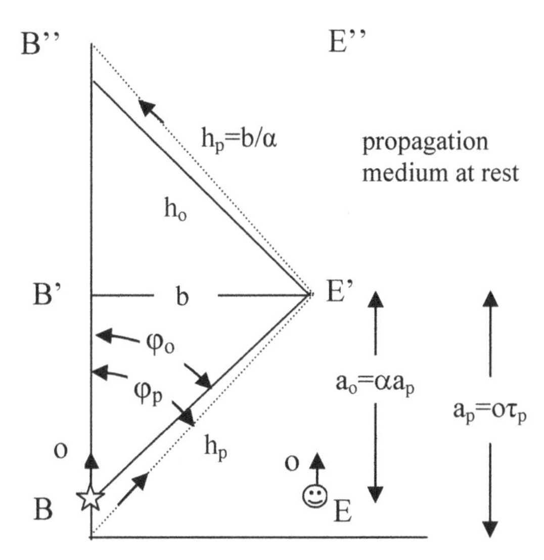

Figure 6.4 Space-space diagram, propagation normal to the direction of motion, fixed source-observer separation

For broadside propagation and no motion $\varphi_p=90^o$ and $a_p=0$. With classical motion, $a_p>0$, and $\varphi_p<90^o$. The source and observer move a distance $a_p=o\tau_p$, in the receding propagation time τ_p. Where τ_p is the time for the wave to propagate a distance h_p along the hypotenuse of the right-angled triangle, in the propagation medium, i.e. the distance between the source emission point and the observer reception position. In terms of the invariant distance b (which does not change being normal to the direction of motion) and equal source and observer Mach numbers M=o/c=s/c, one has:

$$\tau_p = h_p/c, \;\; a_p = o\tau_p = oh_p/c = Mh_p, \;\; M = o/c$$
$$b = (h_p^2 - a_p^2)^{1/2} = h_p(1\text{-}M^2)^{1/2} = h_p\alpha, \;\; \alpha = (1\text{-}M^2)^{1/2} \tag{6.3}$$

Here α is a purely geometrical (classical) effect, nothing to do with relativistic motion. From equations 5.18, Chapter V, $\tau_o = \alpha\tau_p$ and $a_o = \alpha a_p$, where α is now the Lorentz contraction, equation 6.3 becomes:

$$\tau_p = h_p/c = b/\alpha c, \text{ and } \tau_o = \alpha\tau_p = b/c \tag{6.4}$$

Thus the increased propagation time becomes $\tau_p = b/\alpha c$, in the medium with motion. Amazingly, this is offset by the decreased propagation time $\tau_o = \alpha\tau_p$, in the moving frame, *making the receding time in the moving frame $\tau_o = b/c$, independent of motion.* This invariance helped the relativists argument that there was no medium, when in fact the result depends very much on the propagation medium.

Through symmetry, the approaching times τ_{pa}, τ_{oa}, from E' to B'' are the same as the receding times B to E'. Thus the individual one-way propagation times in the moving frame $(OWPT)_o$ given by $\tau_{or} = b/c$, and the round trip propagation time $(RTPT)_o$, given by $\tau_{or} + \tau_{oa} = 2b/c$, are independent of motion. Also, the propagation paths between space ships B-E and B-D in Figure 6.4 are identical. So that signals from B to E and B to D will arrive simultaneously, upholding simultaneity of signals between adjacent ships.

The propagation angles φ_p, φ_o in the stationary and moving frames, where space contracts in the direction of motion, become:

$$\tan\varphi_p = b/a_p = b/Mb/\alpha = \alpha/M, \;\; \tan\varphi_o = b/a_o = 1/M,$$
$$a_p = o\tau_p = Mb/\alpha, \quad a_o = \alpha a_p = Mb \tag{6.5}$$

$$h_o = (b^2 + a_o^2)^{1/2} = b(1+M^2)^{1/2} = (b/\alpha)[(1+M^2)(1-M^2)]^{1/2} = h_p(1-M^4)^{1/2} \tag{6.6}$$

If M=0.6, α=0.8 then φ_p=53° and φ_o=59°, making a 6° difference between angles. For an omni-directional source, the above angles are not obvious or important. However, for ultra-directional sources, such as lasers, these angles would be the only possible angles for communication with position E'. Or, alternatively, to reflect the signal back to position B''.

4.2 In-line propagation

Consider now the communication of one space ship directly ahead of the other, say B-C in the direction of motion. This is illustrated in the space-time diagram in Figure 6.5. The dotted lines indicate the light paths from the source to observer, which will be 45° if t_p is in years and x_p is in light years (LY). It can be seen in the diagram for stationary space ships, indicated by the vertical parallel 'tram lines' B and C (M=0), that the RTPT for the path B-C-B is 2b/c. For the moving ships case, (M≠0), the slopping parallel 'tram lines' give flight paths now with gradients $\tau_p/x_p/c$ =c/s =1/M.

Let the receding propagation distance between the source emission point B and observer reception position C', in the moving frame, be x_{or}. And for position C'' in the stationary propagation medium be x_{pr} with the corresponding propagation times τ_{or} and $\tau_{pr.}$ The distance x_{or} can be expressed in terms of the contracted distance αb and the receding observer distance $o\tau_{or}$ in the moving frame. Thus, one has for a receding observer (moving in the direction of propagation):

$$x_{or} = \alpha b + o\tau_{or}, \quad \text{where} \quad x_{or} = c\,\tau_{or} \qquad (6.7)$$

Rearranging equation 6.7 and dividing by c, gives the receding one way propagation times OWPT for B→C' in the moving observer frame and for BC'' in the stationary propagation medium as:

$$\tau_{or} = b\,c^{-1}\alpha/(1\text{-}M) \quad \text{and} \quad \tau_{pr} = \tau_{or}/\alpha = b\,c^{-1}/(1\text{-}M) \qquad (6.8)$$

Where bc^{-1} is the propagation time for a stationary system, measuring rod or vacuum. Thus the receding propagation time in the moving frame τ_{or} is increased classically by 1/(1-M) and decreased relativistically by α. Let the corresponding approaching path B-A' (moving against the direction of propagation) or return path C'-B' be x_{oa}. Also for the return path C''-B'' or backward path B-A'' in the stationary propagation medium be x_{pa}, with corresponding propagation times τ_{oa} and τ_{pa}, respectfully. The distance x_{oa} can be expressed in terms of the contracted distance αb and the approaching propagation distance $o\tau_{oa}$.

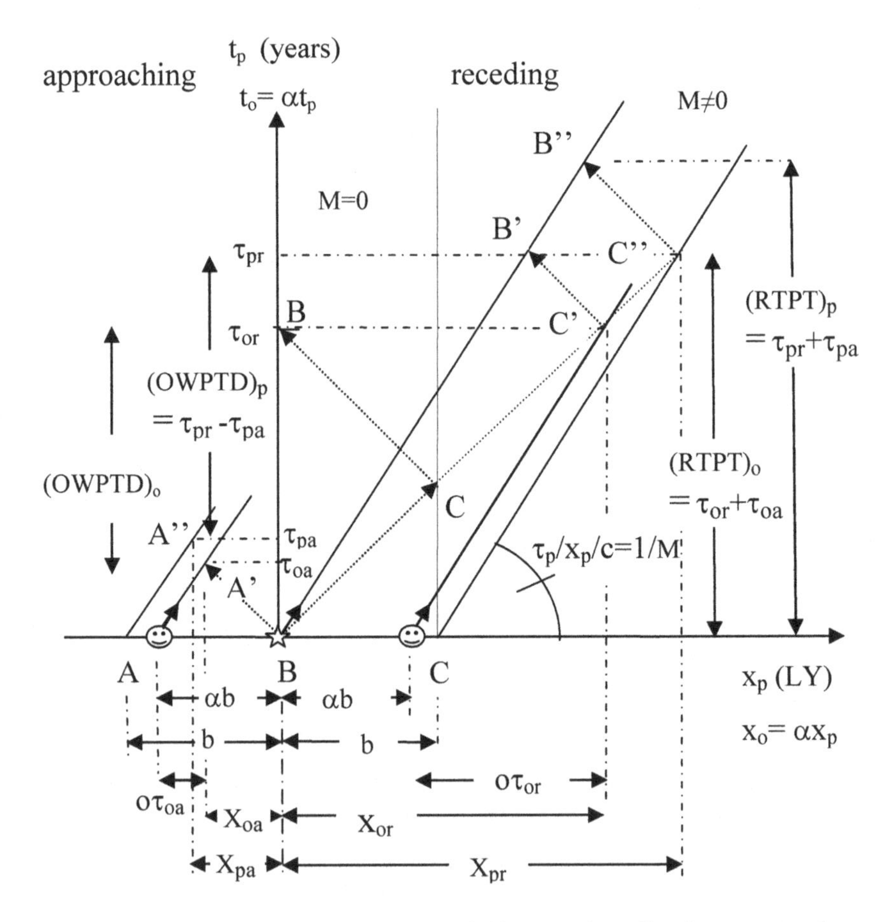

Figure 6.5 Space-time diagram, inline motion, fixed source and observer separation moving relative to the propagation medium

Thus for an approaching observer:

$$x_{oa} = \alpha b - o\tau_{oa}, \quad \text{where} \quad x_{oa} = c\,\tau_{oa} \qquad (6.9)$$

Giving the approaching observer propagation times for C'-B' or B-A' in the moving frame and for C''-B'' or B-A'' in the stationary medium as:

$$\tau_{oa} = b\,c^{-1}\alpha/(1+M) \quad \text{and} \quad \tau_{pa} = \tau_{oa}/\alpha = b\,c^{-1}/(1+M) \qquad (6.10)$$

The approaching propagation time, in the moving frame, is now shortened by both the relativistic α and the classical $1/(1+M)$ terms.

So the total round trip propagation time (RTPT) around the path B-C'-B', in the moving observer frame, and B-C''-B'' in the stationary propagation medium, from equations 6.8 and 6.10, become:

$$(RTPT)_o = \tau_{or} + \tau_{oa} = bc^{-1}\alpha\,[\{1/(1-M)\}+\{1/(1+M)\}]$$

$$= bc^{-1}\alpha 2/\alpha^2 = \alpha^{-1}2b/c \qquad (6.11)$$

$$\text{as} \quad [\{1/(1-M)\}+\{1/(1+M)\}] = 2/\alpha^2$$

$$\text{and} \quad \alpha^2 = (1-M)(1+M) = (1-M^2) \qquad (6.12)$$

$$(RTPT)_p = \tau_{pr} + \tau_{pa} = (RTPT)_o/\alpha = \alpha^{-2}2b/c \qquad (6.13)$$

The term $\alpha^2=(1-M^2)$, in equation 6.12, is not relativistic, it is a purely geometric effect. It is reduced by the relativistic term α, but is not completely neutralized, as in broadside motion, equation 6.4. The time 2b/c can be recognized as the RTPT for stationary source and observer. The RTPT in both the moving frame and stationary medium is a function of frame speed, but is an insensitive method to measure motion, as it is a function of α, which in turn is insensitive to motion at low speeds. However, this is in contrast to the broadside case, where

$(RTPT)_o$=2b/c, in the moving frame, is completely independent of frame speed. The individual receding and approaching one way propagation time (OWPT) between ships B-C' and B-A', given by equations 6.8 and 6.10, respectively, are not simultaneous; they are also dependent on the frame speed. The fractional lengthening and shortening of these times in both the stationary and moving frames is given by the one way propagation time ratio:

$$OWPTR = \tau_{pr}/\tau_{pa} = \tau_{or}/\tau_{oa} = (1+M)/(1-M) \quad (6.14)$$

Equation 6.14 is an important result. It depends only on M and is independent of α and the propagation path length b/c. It therefore provides a sensitive method to measure motion. For $M \ll 1$, OWPTR $\approx 1+2M$, and for change $\Delta OWPTR \approx 2M$, If for example $M=10^{-6}$, $\Delta OWPTR \approx 2.10^{-6}$.

The one-way propagation time difference OWPTD or the difference in 'simultaneity' in the moving observer frames becomes:

$$(OWPTD)_o = \tau_{or} - \tau_{oa} = bc^{-1}\alpha[\{1/(1-M)\}-\{1/(1+M)\}] = M\alpha^{-1}\, 2b/c \quad (6.15)$$

$$\text{as} \quad [\{1/(1-M)\}-\{1/(1+M)\}] = 2M/\alpha^2$$

and in the stationary propagation medium:

$$(OWPTD)_p = \tau_{pr} - \tau_{pa} = (OWPTD)_o/\alpha = M\alpha^{-2}\, 2b/c \quad (6.16)$$

Both OWPTR and OWPTD are functions of frame speed. They are more sensitive to frame speed than the RTPT method, as they have additional M terms, rather than just M^2 in α, which are more sensitive to motion for M<1. OWPTR, equation (6.14), which is a function of M only, could provide a method of measuring the absolute frame speed relative to space, in the form of a cosmic speedometer. Rearranging equation 6.14 one gets:

$$M = (\tau_{or}/\tau_{oa} - 1)/(\tau_{or}/\tau_{oa} + 1) \qquad (6.17)$$

Although there is no effect of motion on propagation time in the broadside position, in the moving frame, the propagation angle φ_o, equation 6.5 is a function of speed.

The parameters used in Figure 6.5 are: $M=s/c=0.6$, $\alpha=[1-(M)^2]^{1/2} = [1-(0.6)^2]^{1/2}=0.8$ and b=1 light year (LY), resulting in the following observable optics. According to equation 6.14, $OWPTR=\tau_{or}/\tau_{oa}= (1+M)/(1-M) = (1+0.6)/(1-0.6) = 4$. According to equation 6.16, the $(OWPTD)_o=\tau_{or}-\tau_{oa}=2bM/\alpha c=2x1cx0.6/0.8xc=1.5$ years. From equation 6.5, $\varphi_p = \tan^{-1}(\alpha/M) = 53°$ and $\varphi_o = \tan^{-1}(1/M) = 59°$. For more realistic speeds such as the Earth through space, $M \approx 0.001$, $\alpha \approx 1$, $OWPTR= \tau_{or}/\tau_{oa}=1.002$ and $(OWPTD)_o= \tau_{or}-\tau_{oa} = 2x1x0.001/1 = 0.002$ years = 61 $x10^3$s for b = 1LY, and $61x10^3/31.5x10^6 \approx 2$ms for b = 1 light seconds (LS). Also $\varphi_o \approx \varphi_p = 89.94°$. (1Y=$31.5x10^6$s, 1LS = $3x10^8$m, 1LY=1LS $x31.5x10^6 \approx 10^{16}$m). These observations are unique for this particular source and observer configuration.

4.3 Individually moving source and observer

It can be seen that because of the presence of the propagation medium, distinction has to be made between the source and observer and which moves relative to the medium. For intricate source and observer flight paths, the observed event time is a complex function of the source time, speed and direction at the source emission point, the observer speed and direction at the observer reception position and the propagation distance between them. Even interchanging the source and the observer, for the same flight paths, will give different observations. As a simple example, Figure 6.6 revisits Einstein's twin paradox, where the source and then the observer make an excursion around the local universe and returns. The source moves first with speed 's' moving in a straight line with respect to the stationary medium and observer. The y and x axes are in years and light years (LY), respectively. Figure (a) illustrates a source making an excursion of Ex=3LY, 6LY there and back, at a motional Mach number M=s/c=0.6. The motion is direct (σ=0) from a stationary observer at the origin in the stationary propagating medium. The observer is moving in time but not in space (vertical line in diagram). The light path communicating the turning point of the source to the observer, indicated by the dotted line, goes from right to left.

The gradient of the source flight path is $c(t_p/x_p)$=c/s=1/M=1.66 or 59° and the gradient of the light path=c/c=1 or 45°, also $\alpha_s=[1-(M)^2]^{1/2}$ =0.8. The source flight path relative to the medium of x_p=3LY, corresponds to a time t_p in the propagating medium ($t_o=t_p$) of t_p= x_p/M=3x1.66=5 years. Now in terms of source event time τ_s (hypotenuse) in the moving frame, from equation 5.21, Chapter V, $\tau_s=\alpha_s\tau_p$= 0.8x5=4 years. Or more precisely from equation 7.62, Chapter VII, τ_s= $(-\tau_p{}^2+(x_p/c)^2)^{1/2}=(-5^2+3^2)^{1/2}=(-25+9)^{1/2}=(-16)^{1/2}$= j4 years, the quantity is in fact imaginary. Note the imaginary triangle ratios 3, 4 and 5. In this kind of geometry, Pythagoras's Theory does not work. The longest side in the right angle triangle is not the hypotenuse, it's the opposite. From Chapter V, equation 5.32, the event time change τ_p in the propagation medium, for a source time change τ_s=8 years (4 years forward, 4 years reverse), becomes for a stationary observer o=0, α_o=1, ε_o=1, $\tau_o=\tau_p$:

$$\tau_p=\left[\left(\frac{\varepsilon_s}{\alpha_s}\right)^+\Delta\tau_s\right]_{\varpi 1}^{\varpi 2}+\left[\left(\frac{\varepsilon_s}{\alpha_s}\right)^-\Delta\tau_s\right]_{\varpi 2}^{\varpi 3}$$

$$=\left[\frac{1+0.6}{0.8}\right]4+\left[\frac{1-0.6}{0.8}\right]4$$

$$=(2)\text{x}(4)+(0.5)\text{x}(4)=8+2=10 \text{ years} \quad (6.18)$$

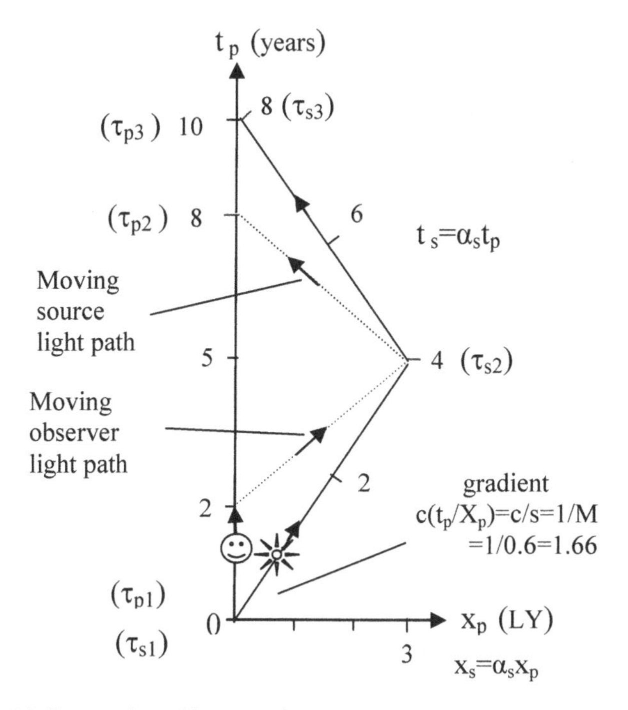

(a) Space-time diagram for source and observer motion

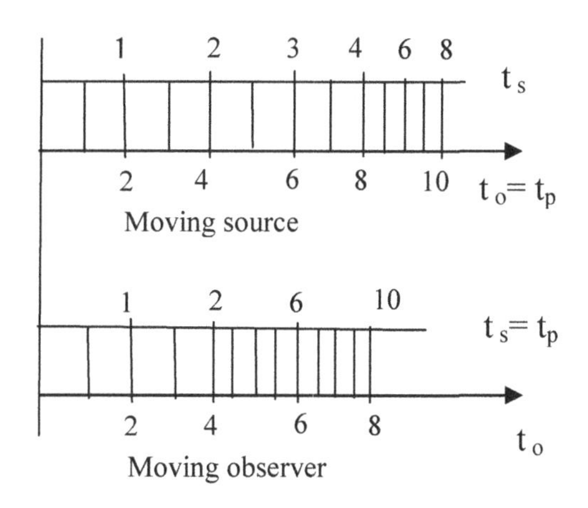

(b) Observed time histories

Figure 6.6 Einstein's twin paradox, M=0.6, α=0.8. There is no paradox when the medium is restored, but there are two situations depending on whether the source or observer moves

Thus τ_s= 4 and 8 years on the right side of Figure 6.6 (a) become, from equation 6.18, τ_p= 8, 10 years on the left-hand side of the figure (4 years expanded into 8 and then 4 years compressed into 2. These times are compared in Figure (b). The corresponding propagation frame (τ_p=10yrs, x_p=6LY) becomes in the moving frame ($\tau_s=\alpha_s\tau_p$=0.8x10=8 years, $x_s=\alpha_s x_p$=0.8x6=4.8 LY), i.e. the same distance is covered in less time in the moving frame, i.e. the astronauts get there quicker in the moving frame. Across frames, the hybrid Mach number, from equation 5.24, becomes $M^*=x_p/\tau_s c=x_p/\tau_p\alpha_s c=M/\alpha_s$= 0.6/0.8=0.75, i.e. higher than the Mach number in the medium. This allows the astronaut to travel vast distances in the familiar world, in his/hers slower time, or effective higher speed.

Also shown in Figure (b), the source and observer are interchanged for the same flight path, i.e. for a stationary source, and now the observer makes the same round trip. In a similar analysis where the light path now runs from left to right, it can be seen that the observed time is now different. In spite of Einstein's claim that only relative motion between systems is important the observed time histories are quite different depending on whether the source or observer moves relative to the medium.

4.4 Source and observer moving together

A slightly more complicated example is illustrated in Figure 6.7. Here both the source and observer (observer is now on the right), separated by a fixed distance b of 1.0 LY, oscillate (move forward and backward) together relative to the propagating medium, again over a 6 LY excursion at a M=0.6, α=0.8. The contracted distance in the direction of motion is αb=0.8 LY. The light paths now go from left to right. From equation 6.8, τ_{pr}=b/(1-M)=1.0 /(1-0.6)=2.5 years. i.e. there is a 2.5 year propagation time delay in the stationary medium, which becomes $t_o=\alpha_o t_p$ =0.8x2.5=2.0 years, in the moving frame, before the event is seen.

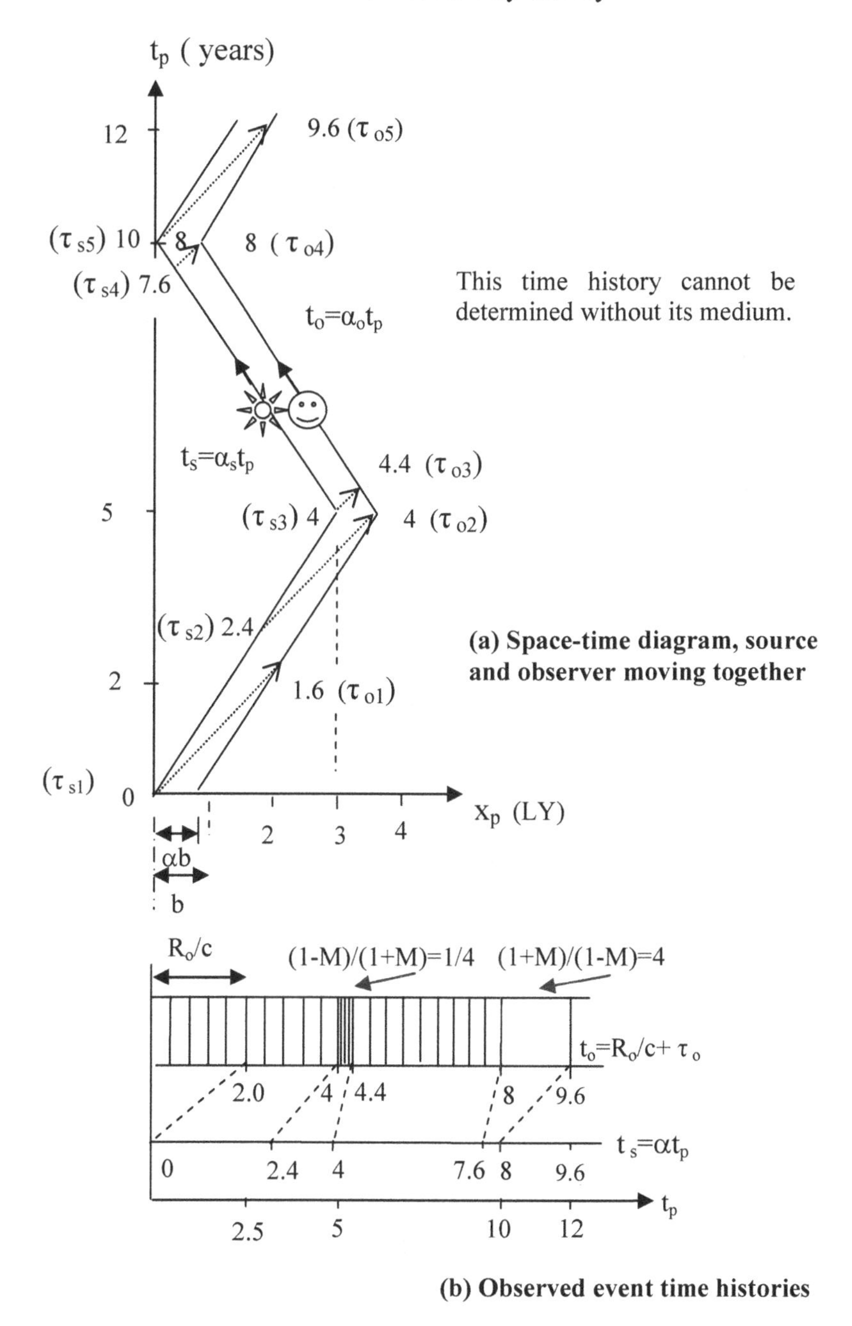

(b) Observed event time histories

Figure 6.7 Oscillating source and observer of fixed separation, b=1LY, M=0.6, α=0.8, oscillating period 10 years. There is no relative motion between systems, only constant motion and reversing

The source event time intervals τ_{s2}, τ_{s1} etc., in between significant events (determined by the intersection of the light paths and the source and observer flight paths), in the moving frame, are τ_s=0, 2.4, 4, 7.6 and 8 years, shown on the left of the parallel 'tram' lines. This gives incremental difference times of $\Delta\tau_s$ = 2.4, 1.6, 3.6 and 0.4 years. As the source and observer are moving at the same speed α_s=α_o, this gives incremental observer event times t_o on the right of the tram lines according to equation 5.32, Chapter V as:

$$\Delta t_o = R_o/c, \left[\frac{\varepsilon_s}{\varepsilon_o}\Delta\tau_s\right]_{\varpi 1}^{\varpi 2}, \left[\frac{\varepsilon_s}{\varepsilon_o}\Delta\tau_s\right]_{\varpi 2}^{\varpi 3}, \left[\frac{\varepsilon_s}{\varepsilon_o}\Delta\tau_s\right]_{\varpi 3}^{\varpi 4}, \left[\frac{\varepsilon_s}{\varepsilon_o}\Delta\tau_s\right]_{\varpi 4}^{\varpi 5} \quad (6.19)$$

$$= R_o/c, \left[\frac{1-M}{1-M}\right]2.4, \left[\frac{1-M}{1+M}\right]1.6, \left[\frac{1+M}{1+M}\right]3.6, \left[\frac{1+M}{1-M}\right]0.4 \quad (6.20)$$

$$\Delta t_o = R_o/c,\ 2.4, \left[\frac{0.4}{1.6}\right]1.6,\ 3.6, \left[\frac{1.6}{0.4}\right]0.4 = 1.6,\ 2.4,\ 0.4,\ 3.6,\ 1.6 \quad (6.21)$$

$$\text{Accumulative } t_o = 1.6,\ 4,\ 4.4,\ 8, \text{ and } 9.6. \quad (6.22)$$

The individual motional event time ratios ($\Delta\tau_o/\Delta\tau_s$), over the cycle, now become: 2.4/2.4=1, 0.4/1.6=1/4, 3.6/3.6=1 and 1.6/0.4=4 respectively. Or from equation 6.14, for M=±0.6, the one way propagation time ratios OWPTR=(1-M)/(1+M)=1/4 (and 4) i.e. a maximum time compression and expansion of four times the stationary value will occur. The three event time scales t_p, t_s and t_o are compared in Figure 6.7 (b).

According to Einstein's Inertial Frame (no propagation medium, no propagation time change with motion) Section 1.3, Chapter I, no propagation time change can occur for constant motion, at any speed, between a source and observer of fixed separation. Thus, from Einstein's simultaneity, upstream τ_{s2}-τ_{o2} and downstream τ_{s3}-τ_{o3} propagation times (distances), in Figure 6.7, should be equal, making the above ratios unity, with or without motion. This is clearly not the case, Einstein's invariant propagation time in the moving frame is not in accord with Lorentz or Sagnac. For those relativists, not convinced of the presence of the propagation medium, the invalidity of Einstein's relativity can be checked by running this thought experiment in reality.

5 Measurement Methods

5.1 Available measurements.

Sagnac demonstrated that the medium existed and was stationary with respect to the Earth through his rotating mirrors experiment, described in Section 5.1 and Figure 4.4(a) in Chapter IV. It was shown that his optical loop (of circumferential distance d) generated a time delay, relative to the stationary medium, given by $\Delta t=\Delta d/c$, where $\Delta d=vt=vd/c$, thus $\Delta t=vd/c^2=Md/c$. This classical measurement is based directly on $M=v/c$. It is an efficient (sensitivity), practical one way velocity measurement. Other methods considered in this chapter, where b is the source/observer fixed separation distance, 2b/c is the stationary round trip propagation time, For speed on Earth, α has negligible effect and assuming $M<<1$, one has:

i) Round trip propagation time in the moving frame $(RTPT)_o=\alpha^{-1}(2b/c)\approx(1+M^2/2)(2b/c)$, equation 6.11 and $\Delta(RTPT)_o\approx(M^2b/c)$.
ii) One way propagation time $(OWPT)_o=\alpha(1-M)^{-1}(b/c)\approx(1+M)(b/c)$, equation 6.8 where $\alpha\approx1$, and $\Delta(OWPT)_o\approx Mb/c$.

iii) One way propagation time ratio OWPTR=(1+M)/(1-M)≈ 1+2M, equation 6.14 and ΔOWPTR≈2M.

iv) One way propagation time difference $(OWPTD)_o=M\alpha^{-1}(2b/c)$, equation 6.15 and $\Delta\,(OWPTD)_o \approx M2b/c$.

v) Sagnac's optical loop delay, equation 4.4, Δt=Md/c, d=b.

The first item $(RTPT)_o$ is the least sensitive to speed, as it depends on M^2, which is a very weak function of M at low speeds. The early interferometer experiments used this insensitive method, including the MMX (1887). For the Earth orbiting around the Sun, $M \approx 10^{-4}$, and b=3m, the $RTPT=M^2b/c \approx 10^{-8}x3/3x10^8 \approx 10^{-16}s$. Using a laser interferometer of wavelength λ=600nm= $6x10^{-7}m$, gives a fringe displacement number of $N=cRTPT/\lambda=3x10^8x10^{-16}/6x10^{-7}=1/20$, which is just about discernable. The remaining methods contain M, which can measure orders of magnitude lower speeds. For practical speeds on Earth $M \approx 10^{-6}$, $\Delta t \approx Mb/c=10^{-6}x3/3x10^8 \approx 10^{-14}s$ giving $N=c\Delta t/\lambda=Mb/\lambda= 10^{-6}\,x3x6^{-1}x10^7=5$. For M&G (1925), Section 5.3 and 5.4, equation 4.9, Chapter IV, $\Delta t \approx dhM/cR$, if d=h=1.4 km, $M=10^{-6}$ and $R=6.4x10^6$, $\Delta t \approx 2x10^6\,x10^{-6}/(3\,x10^8x6.4x10^6) \approx 10^{-15}s$ or N=0.5. The one way system is sensitive enough to investigate medium profiles on and around and between planets.

5.2 Rotary motion on Earth

For a simple classical time delay experiment on Earth, a transmitter could be rotated on a rotor arm and fed through slip rings from a stationary signal generator. The transmitter output could be received and compared directly to the stationary generator output, through a signal comparator or oscilloscope, as illustrated in Figure 6.8. For a rotor rotating say at the speed of sound, 320m/s, i.e. $M \approx 10^{-6}$, a classical time modulation ratio of OWPTR = (1+M)/(1-M) ≈ 1+2M = 1.000002

or a change of Δ(OWPTR) ≈ 2M = 2.10^{-6} will occur. The transmitter could be driven, for example, by a source frequency of say f_s=100

MHz (10^8) from a stationary signal generator. This would give a frequency modulation of Δ(OWPTR) x f_s $=2.10^{-6}x10^8=200$ Hz (and 25Hz) at the signal comparator. If there is no medium there would be no frequency modulation. If the system moved sideways relative to the medium, the modulation would be asymmetrical, revealing its motion.

5.3 Relativistic measurements

The time contraction in a moving source frame τ_s compared to a stationary medium time τ_p depends on $\alpha=[1-(M)^2]^{1/2}$. The moving source relativistic time change then becomes $\Delta\tau_s=\Delta\alpha\tau_p$, where for small M, $\Delta\alpha \approx -(1/2)M^2$. For a source in linear motion or rotating on a rotor arm, as in Figure 6.8, again at approximately the speed of sound, $M\approx 10^{-6}$, $\Delta\alpha \approx -(1/2)M^2 \approx -0.5x10^{-12}$. A signal generator could be used as before. Or clocks could be used, one stationary the other moving. Or one clock stationary and its time transferred electronically to the moving frame. Clocks with sufficient absolute stability, variability smaller than 10^{-12}, are available to measure these small changes. Such clocks are for example Mossbauer Fe^{57}, with stability better than 10^{-13}.

Now for say a 24 hour measurement period i.e. $\tau_p=8.6x10^4$ seconds, the moving source time slows relative to the Earth's time by $\tau_s=\Delta\alpha\tau_p\approx -0.5x10^{-12}x8.6x10^4 \approx -4.3.10^{-8}$ or 43ns lag. Or if a source frequency f_s of 100 MHz is used, a phase delay of $\Delta f_s=\Delta\tau_s f_s=4.3x10^{-8}x10^8=4.3$ Hz, over a 24 hour period, should be obtained well within measurement accuracy. The dependence on velocity rather than acceleration, can be established by varying the rotor diameter for the same tip speed. The classical effect depends on the measurement angle with the axis of rotation. Whereas, the relativistic effect is independent of position.

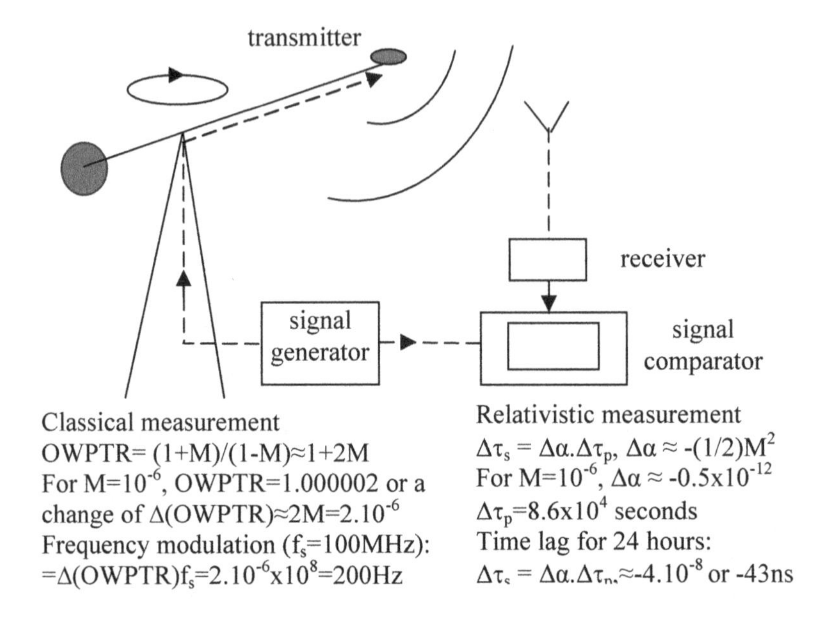

Figure 6.8 Classical and relativistic measurement of motion relative to the propagation medium

6 Conclusions

The effect of source and observer motion on the source event time is established. It is possible, causally, to visually travel to the past and return, but not to interfere with past events. However, it is against causality to move materially into the future or past. The propagation time change through classical wave dynamics, and relativistic time ageing, have been investigated, both perpendicular and in the direction of motion. Worked examples are given and experiments described to obtain further details of the propagation medium on Earth, around planets and in free space.

Chapter VII:

Graphical Extension

1 Introduction

Chapter V analysed sources and observers in motion, with respect to the propagation medium, through solving the EM wave equation. This chapter deals with the graphical extension of the Lorentz Transform (LT), for both source and observer motion relative to the medium. The relativistic addition of velocities is considered and some well known propagation path examples analysed. This includes reflections from a moving train window and from a mirror on an adjacent wall. Bradley's stellar aberration and light convection by a moving more massive medium than a vacuum. Comparison is made with Einstein's special and.

2 Sources and Observers

Accepting the presence of the propagation medium, means that more precision is required to distinguish between the stationary propagation medium, moving frames, emission and reception positions, whether the source or observer moves relative to the medium and whether the source or observer is approaching or receding. This complete situation is illustrated in the space-time diagram in Figure 7.1, which is an extension of Figure 3.2 in Chapter III.

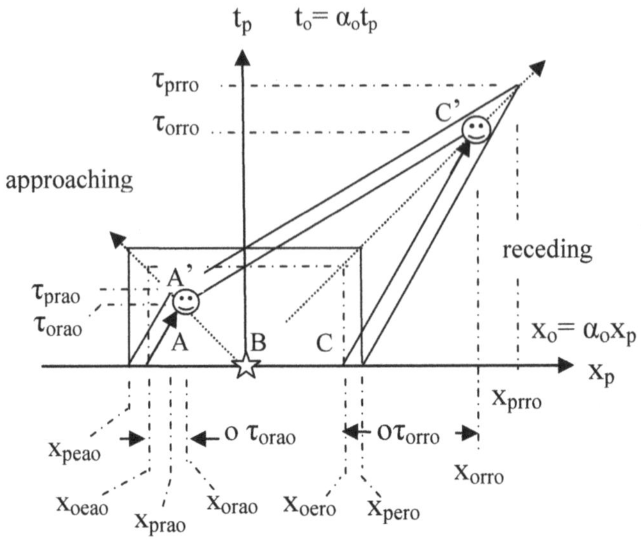

(a) Approaching and Receding Observer

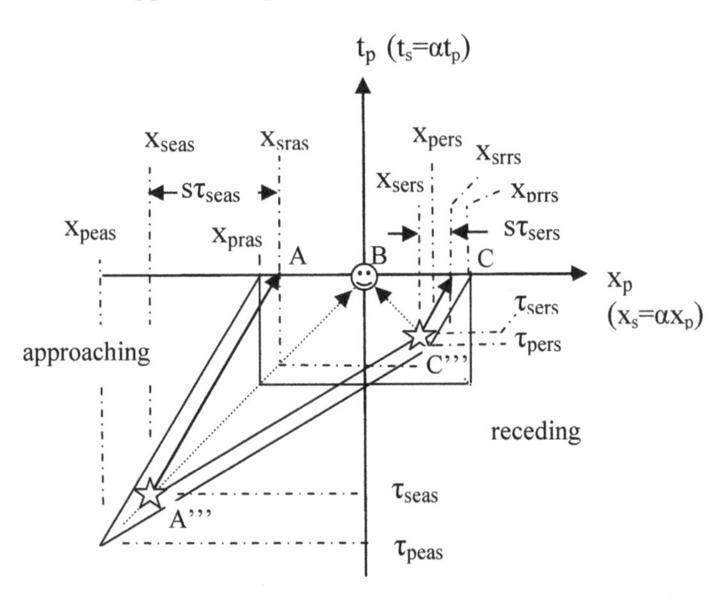

(b) Approaching and Receding Source

Figure 7.1 Lorentz extension for source and observer motion relative to a propagation medium, graphical transform

Here $M_o=o/c$, $\alpha_o=(1-M_o^2)^{1/2}$, $M_s=s/c$, and $\alpha_s=(1-M_s^2)^{1/2}$, also $x=c\tau$, $\tau_o=\alpha_o\tau_p$, $x_o=\alpha_o x_p$, $\tau_s=\alpha_s\tau_p$ and $x_s=\alpha_s x_p$. The subscripts o, s and p indicate moving observer, moving source and the stationary propagation medium, respectively. The subscripts e and r indicate emission and reception positions. The order of the subscripts are: 1st- moving observer/source frame or stationary propagation frame, 2nd- emission or reception position, 3rd- approaching or receding system, 4th- source or observer moving.

These distinctions are necessary to emphasise the various situations, they are simplified later. It can be seen from Figure 7.1 that for an:

Approaching Observer

$$x_{oeao} = x_{orao} + o\ \tau_{orao} \tag{7.1}$$

$$c\ \tau_{oeao} = c\ \tau_{orao} + o\ \tau_{orao} \tag{7.2}$$

$$\tau_{oeao} = \alpha_o \tau_{peao} = \tau_{orao}(1+M_o) = \tau_{orao} + o\ x_{orao}/c^2 \tag{7.3}$$

$$\tau_{orao} = \tau_{peao}\alpha_o/(1+M_o) = \tau_{oeao}/(1+M_o) \tag{7.4}$$

Receding Observer

$$x_{orro} = x_{oero} + o\ \tau_{orro} \tag{7.5}$$

$$c\ \tau_{orro} = c\ \tau_{oero} + o\ \tau_{orro} \tag{7.6}$$

$$\tau_{oero} = \alpha_o \tau_{pero} = \tau_{orro}(1-M_o) = \tau_{orro} - o\ x_{orro}/c^2 \tag{7.7}$$

$$\tau_{orro} = \tau_{pero}\alpha_o/(1-M_o) = \tau_{oero}/(1-M_o) \tag{7.8}$$

The last part of equations 7.3 and 7.7 are the propagation times between a source and observer, they depend on the observer speed relative to the medium. Equation 7.4 and 7.8 are the Lorentz contracted Doppler effect on the observed event time.

Approaching Source

$$x_{seas} = x_{sras} + s\ \tau_{seas} \quad (7.9)$$
$$c\ \tau_{sea} = c\ \tau_{sras} + s\ \tau_{seas} \quad (7.10)$$
$$\tau_{sras} = \alpha_s \tau_{pras} = \tau_{seas}(1 - M_s)$$
$$= \tau_{seas} - o\ x_{seas}/c^2 \quad (7.11)$$
$$\tau_{pras} = \tau_{seas}(1 - M_s)/\alpha_s$$
$$= \tau_{peas}(1 - M_s) \quad (7.12)$$

Receding Source

$$x_{srrs} = x_{sers} + s\ \tau_{sers} \quad (7.13)$$
$$c\ \tau_{srrs} = c\ \tau_{sers} + s\ \tau_{sers} \quad (7.14)$$
$$\tau_{srrs} = \alpha_s\ \tau_{prrs} = \tau_{sers}(1 + M_s)$$
$$= \tau_{sers} + s\ x_{sers}/c^2 \quad (7.15)$$
$$\tau_{prrs} = \tau_{sers}(1 + M_s)/\alpha_s$$
$$= \tau_{pers}(1 + M_s) \quad (7.16)$$

These sets of equations are an extended form of the LT. The first two sets are for an approaching (A') and receding (C') observer with a stationary source (B). The second two sets are for an approaching (A''') and receding source (C''') with a stationary observer (B). Equations 7.4, 7.8, 7.12, and 7.16, are expressed in a classical Doppler time delay ($1 \pm M$) and relativistic contracted (α) forms. Equation 7.7 can be recognized as the classical Lorentz transform. This equation describes a receding observer - stationary source situation. It is only one equation in the complete set above that completes the general LT for source and observer motion relative to the propagation medium.

These results are for direct (inline) source and observer motion. Assuming the Optical Principle of Relativity, where an approaching observer-stationary source and approaching source-stationary observer have similar appearances i.e. they have the same event time ratios.

Then from equations 7.4 and 7.12, the Lorentz contraction α is obtained from:

$$\tau_{orao}/\tau_{peao} = \tau_{pras}/\tau_{seas}$$

giving (7.17)

$$\alpha_o(1+M_o)^{-1} = (1-M_s)\alpha_s^{-1}$$

Where for equal speeds $M_o=M_s=M$ and $\alpha_o=\alpha_s=\alpha$, we then have:

$$\alpha= [(1+M)(1-M)]^{1/2}= (1-M^2)^{1/2} \qquad (7.18)$$

In the case of constant direct motion, the event time transforms K_t, in equation 6.2, Section 2, Chapter VI, is the same for similar equal source and observer motions. This gives some credence to Einstein's claim that only relative motion is important. However, the propagation distances between the source emission and observer reception positions are not equal for equal initial separation distances. The propagation distance for an approaching observer x_{orao}, is shorter than an approaching source x_{seas}. Conversely, a receding observer propagation distance x_{orro} is larger than a receding source propagation distance x_{sers}. Comparing the propagation distances in Figures 7.1(a) and 7.1(b), it can be seen that:

$$x_{orao} < x_{seas} \text{ and } x_{orro} > x_{sers} \qquad (7.19)$$

Thus through unequal propagation distances, directly approaching observers and sources are not identical optically, neither are directly receding observers and sources. Thus interchanging the source and observer, for a fixed separation, in a moving frame will change the propagation distance. For non direct approaches, $\sigma \neq 0$, the classical term becomes directional i.e. $(1 \pm M) \rightarrow (1 \pm M_c os\sigma)$. The relativistic term α remains omni-directional, causing the event times now to be different between source and observer motion. Also, identical closed loop flight paths for individual source and observer motion, will give different observations, as discussed in Section 4.3, Chapter VI.

Therefore, for arbitrary source and observer motion, in general, one can conclude that source and observer motion is not identical or interchangeable, from an optical point of view.

3 Relativistic Addition of Velocities

In this revised theory, there can be motion only between a moving observer 'o' or source 's' and the stationary propagation medium 'p', i.e. relative motion directly between s and o has no fundamental meaning in predicting observations. Although Relativistic Addition of Velocities (RAV) has been used in the past, to attempt to support relative motion, examination shows that it cannot, as it is itself based on the LT (receding observer, stationary source), and Lorentz's propagation medium. It should be said at the outset that RAV, although beautiful mathematically, it is not a very intuitive topic, and if one is not very mathematically inclined they should probably move on to Section 4.

3.1 X direction

Using the receding observer case in Figure 7.1(a) and removing subscripts, except for the moving observer o, stationary propagation medium p and temporarily e for emission and r for reception, one has in the stationary propagation medium in the direction of motion:

$$u_{px} = dx_p / d\tau_p \qquad (7.20)$$

In terms of the observer time τ_o and position x_o in the moving frame, from equations 7.5 one has:

$$x_{or} = x_{oe} + o\tau_{or}, \quad x_{oe} = \alpha\, x_{pe}$$

Rearranging and removing emission and reception subscripts:

$$x_p = (x_o - o\tau_o)/\alpha \quad (7.21)$$

From equation 7.7:

$$\tau_{oe} = \tau_{or} - ox_{or}/c^2, \quad \tau_{oe} = \alpha\, \tau_{pe}$$

giving

$$\tau_p = (\tau_o - ox_o/c^2)/\alpha \quad (7.22)$$

Differentiating equations 7.21 and 7.22 with respect to τ_o one obtains, respectively:

$$dx_p/d\tau_o = (u_{ox} - o)/\alpha, \quad u_{ox} = x_o/\tau_o \quad (7.23)$$

and

$$d\tau_p/d\tau_o = (1 - ou_{ox}/c^2)/\alpha \quad (7.24)$$

Thus the velocity in the stationary propagation medium u_{px}, from equation 7.20, now becomes in terms of velocity u_{ox} in the moving observer frame, using equations 7.23 and 7.24:

$$u_{px} = dx_p/d\tau_p = (dx_p/d\tau_o)(d\tau_o/d\tau_p) = (u_{ox} - o)/(1 - ou_{ox}/c^2) \qquad (7.25)$$

or rearranging 7.25 in terms of u_{ox}

$$u_{ox} = (u_{px} + o)/(1 + u_{px}M/c), \quad M = o/c \quad (7.26)$$

It is interesting to note that equations 7.25 and 7.26 are independent of the Lorentz contraction α.

3.2 Y direction

For the y direction, there is both an x and y contribution, thus:

$$u_{py} = dy_p/d\tau_p = (dy_p/d\tau_o)(d\tau_o/d\tau_p) = u_{oy}(d\tau_o/d\tau_p) \qquad (7.27)$$

as

$$dy_p/d\tau_o = dy_o/d\tau_o = u_{oy}$$

where $y_p = y_o$ (no contraction perpendicular to motion)

From equations 7.27 and 7.24

$$(d\tau_p/d\tau_o) = u_{oy}/u_{py} = (1 - ou_{ox}/c^2)/\alpha \qquad (7.28)$$

$$\text{or} \qquad u_{oy} = u_{py}(1 - u_{ox} M/c)/\alpha \qquad (7.29)$$

Substituting for u_{ox} in equation 7.29, using equation 7.26, and after some rearranging one has:

$$u_{oy} = u_{py}\,\alpha/(1 + u_{px}M/c) \qquad (7.30)$$

Here, at right angles to motion, equations 7.29 and 7.30 are now functions of the Lorentz contraction α.

3.3 Moving frames

1. For light moving parallel with the x-axis in the stationary frame i.e. u_{px}=c, its appearance in the x direction in the moving frame is from equation 7.26, $u_{ox} = c_x = (c+o)/(1+o/c) = c$.
2. For light moving parallel with the x-axis in the moving frame i.e. u_{ox}=c, its appearance in the x direction in the stationary frame is from equation 7.25, $u_{px} = c_x = (c-o)/(1-o/c) = c$.
3. For light moving parallel with the y-axis in the stationary frame i.e. u_{py}=c and u_{px}=0 (zero), its appearance in the y direction in the moving frame is from equation 7.30, $u_{oy} = \alpha c$.
4. For light moving parallel with the y-axis in the moving frame i.e. u_{oy}=c and u_{ox}=0 (zero), its appearance in the y direction in the stationary frame is from equation 7.29, $u_{py} = \alpha c$.

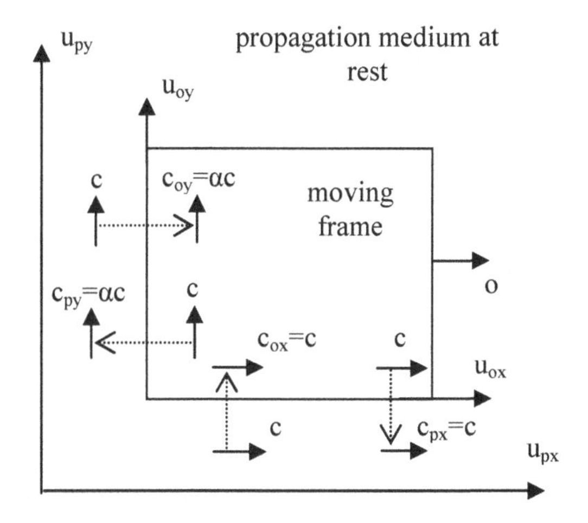

Figure 7.2 X and y velocity transforms between frame motion with velocity 'o' relative to the propagation medium. The propagation speed c is invariant across the x axes, but is a function of α (frame speed) across the y axes

In summary, as illustrated in Figure 7.2, the propagation speed 'c' in the x direction (direction of motion) is invariant, going from the stationary propagation medium to a moving frame and vice versa. This gives an impression of no difference between a moving and stationary frame. However, the propagation time in the x direction, in the moving frame, from equation 7.8 becomes $[\alpha_o/(1\text{-}M_o)]x/c$, i.e. it is a function of frame speed, revealing its motion, also illustrated in Figure 4.1

Although, the propagation speed in the y direction (perpendicular to motion), across frames, is variant (αc). The propagation time in the y direction, in the moving frame, becomes $t_o=\alpha t_r=\alpha x/\alpha c=x/c$, which is independent of frame speed, it does not reveal motion.

4 Medium Paths

The actual light paths in the propagation medium are now identified. Firstly, light appearing to be propagating in a moving frame is analyzed in terms of its actual path within the stationary propagation medium. Secondly, light propagating in the stationary propagating medium is analyzed in terms of its appearance in a moving observer frame. Einstein's railway carriage thought experiment is considered first and extended.

4.1 Window reflection (propagation perpendicular to motion)

The relativists position, based on Einstein's Inertial Frame (EIF), without a propagation medium, is illustrated in Figure 7.3(a). It shows a source and observer of fixed separation distance b, aligned perpendicular to the direction of motion, travelling at an observer speed 'o'. The chain dotted line indicates the apparent propagation path, in an unknown propagation system, based on relative motion and no medium. The propagation time in the moving frame, is then given by:

$$\tau_o = b/c. \qquad (7.31)$$

As b and c are invariant, the propagation time τ_o in the moving frame is independent of frame speed. This model gives the right result but apparently for the wrong reason.

According to the revised theory, the propagation path b and the propagation angle $\theta_o = 90^o$ in Figure 7.3 (a) are false. Propagation actually takes place in the stationary propagation medium. If a laser were placed at the source position and fired directly at the observer, the light would miss it, being blown to the left by the medium 'side wind.' For the light to be received at the observer, via the propagation medium, it would need to be pointed to the right of the observer.

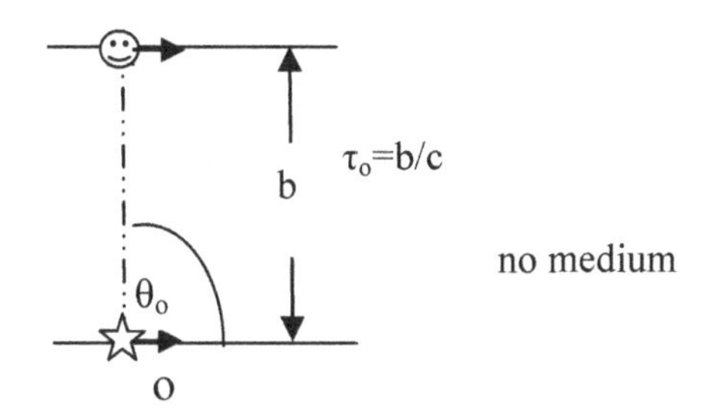

(a) Ether-less theory-relativity (moving frame)

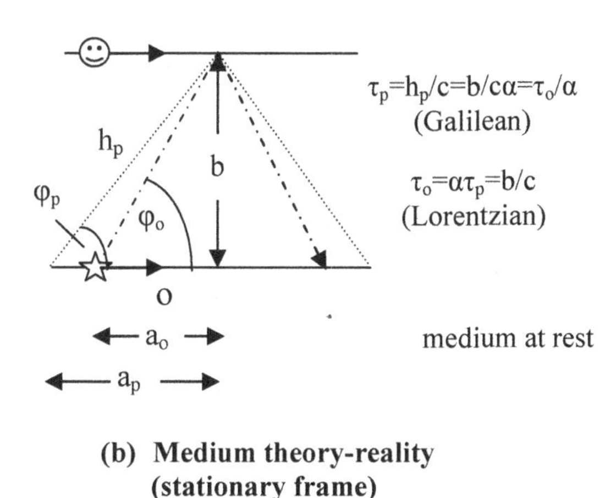

(b) Medium theory-reality (stationary frame)

Figure 7.3 Reflection from window aligned with direction of motion

Figure 7.3(b) represents the reflection from the window of a moving carriage. The actual light propagation path in the medium for the reflection to occur (that which would be seen in the stationary frame with chalk dust and a laser beam), is given by the dotted line in the figure. The chain dotted line is the light path in the medium that would be seen by the moving observer. For the isosceles triangle, where b is unchanged through being at right angles to the direction of motion, one has for the propagation time τ_p in the stationary medium:

$$\tau_p = h_p/c, \;\; a_p = o\tau_p = oh_p/c = Mh_p, \; M = o/c \quad (7.32)$$

$$b = (h_p{}^2\text{-}a_p{}^2)^{1/2} = (h_p{}^2\text{-}M^2h_p{}^2)^{1/2} = h_p\alpha, \;\; \alpha = (1\text{-}M^2)^{1/2} \quad (7.33)$$

$$\text{giving} \quad \tau_p = h_p/c = b/\alpha c, \;\text{ and }\; \tau_o = \alpha\tau_p = b/c \quad (7.34)$$

The first part of equation 7.34 τ_p=b/αc is Galilean i.e. classical propagation in the stationary medium, where α is a geometric term not relativistic (Lorentzian). Fortunately, the second term in equation 7.34, which is relativistic, i.e. τ_o=$\alpha\tau_p$, neutralizes the geometric term, maintaining the invariant propagation time τ_o=b/c, which supports the relativist's view. *However, if chalk dust and a moving laser were used, there would be no laser beam perpendicular to the motion, it would lie at an angle φ_p to the direction of motion in the medium.* With contraction in direction of motion in the moving frame, x_o=αx_p, one has:

$$a_o = \alpha a_p = \alpha Mh_p = Mb = ob/c = o\tau_o \quad (7.35)$$

$$\tan\varphi_p = b/a_p = \alpha/M \quad (7.36),$$

$$\tan\varphi_o = b/a_o = 1/M \quad (7.37)$$

From equations 7.36, the propagation path subtends an angle φ_p= $\tan^{-1}(\alpha/M)$ in the propagation medium. At the moving observer, the propagation angle will be φ_o= $\tan^{-1}(1/M)$. Again in the moving frame the geometric α is neutralized by the relativistic α.

Using the formal Relativistic Addition of Velocities (RAV) a stationary point in the x direction, u_{ox}=0, in a frame moving at velocity 'o' with respect to the medium, becomes in the stationary medium, from equation 7.25 u_{px}=(u_{ox}-o)/(1-ou_{ox}/c^2)=-o. A light ray parallel with the y axis (u_{oy}=c, u_{ox}=0), in the moving frame, appears in the y direction in the stationary medium, from equation 7.28 u_{py}=$u_{oy}\alpha$/(1-ou_{ox}/c^2)

=αc. Thus transforming from the moving to the stationary medium frame, using RAV, one has:

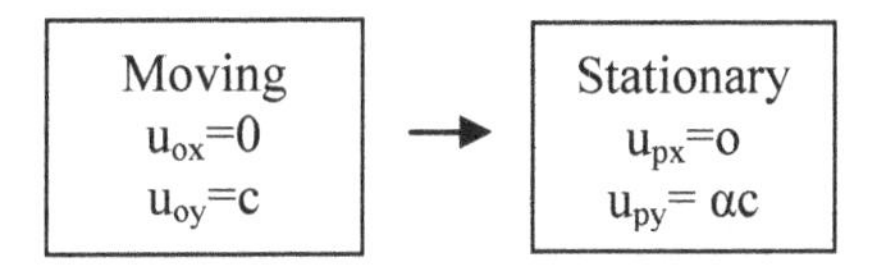

This confirms the propagation angles in the moving and stationary frames as:

$$\tan\theta_o = u_{oy}/u_{ox} = c/0 = \infty,\ \ \theta_o = 90^o \quad (7.38)$$

and

$$\tan\varphi_p = u_{py}/u_{px} = \alpha c/o = \alpha/M \quad (7.39)$$

Note: θ_o is a RAV angle in the moving frame normal to the direction of motion. Whereas φ_o is the observer angle in the medium in the moving frame, both are illustrated in Figure 7.3. Thus equation 7.31, Figure 7.3(a), gives the right result but for the wrong reason. The actual propagation situation is illustrated in Figure 7.3(b).

4.2 Mirror reflection (propagation aligned with motion)

Consider now the reflection from a mirror on a wall perpendicular to the carriage motion. Here αb is now the reduced distance to the mirror in the moving frame. Figure 7.4 illustrates three stages of reflection, where subscripts are defined in Figure 7.1. (1) initial separation distance at emission, (2) position at reflection and (3) position at reception. From the figure and equation 7.5, where x_{oero}= αb is for the receding mirror up to the point of reflection (position 2):

$$x_{orro} - o\tau_{orro} = \alpha b \quad (7.40)$$

giving

$$c\tau_{orro} - o\tau_{orro} = \alpha b \quad (7.41)$$

$$\tau_{orro}(1 - M_o) = \alpha b/c \quad (7.42)$$

$$\tau_{orro} = \alpha(1\text{-}M_o)^{-1}b/c \qquad (7.43)$$

For reflection from the mirror to the reception of the approaching observer (position 3), from equation 7.1 where $x_{oeao}= \alpha b$:

$$x_{orao}+ o\tau_{orao} = \alpha b \qquad (7.44)$$

giving

$$c\,\tau_{orao}+ o\tau_{orao} = \alpha b \qquad (7.45)$$

$$\tau_{orao}(1+M_o) = \alpha b/c \qquad (7.46)$$

$$\tau_{oroa} = \alpha\,(1+M_o)^{-1}\,b/c \qquad (7.47)$$

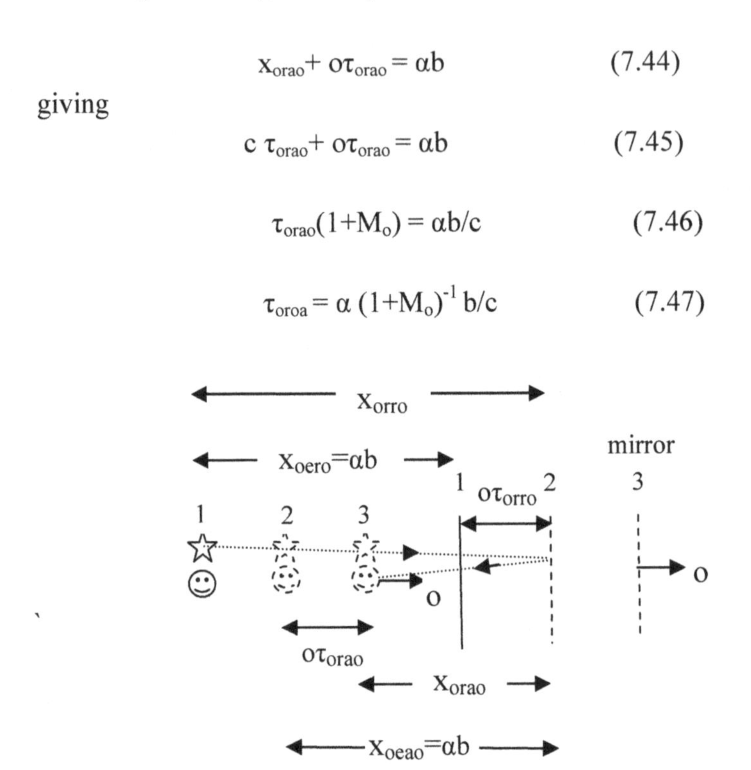

Figure 7.4 Reflection from mirror normal to direction of motion

The total round trip propagation time in the moving frame is then:

$$(RTPT)_o = \tau_{orro} + \tau_{orao} = bc^{-1}\alpha\,[\{1/(1\text{-}M)\}+\{1/(1+M)\}]$$

$$= bc^{-1}\alpha\,2/\alpha^2 = 2b/c\alpha \qquad (7.\,48)$$

where we have used $[\{1/(1\text{-}M)\}+\{1/(1+M)\}] = 2/\alpha^2$

2b/c is now the round trip propagation time (RTPT) for a stationary source and observer. Thus, unlike the reflection from the carriage window, it can be seen that the round trip propagation time, for in-line motion (reflection from the mirror), increases with speed by α^{-1}. This variable propagation time delay, in the direction of motion, illustrated in Figure 7.4(a), does not support the relativists invariant point of view. The variance is in accord with Lorentz, it can be confirmed, using Sagnac's optical loop rotating relative to the stationary medium on Earth, Section 5.1, Chapter IV.

4.3 Bradley's stellar aberration

According to Bradley (1725), an observer (telescope) stationary on Earth, orbiting around the Sun, needs to be pointing slightly in front of a star to be observed. The effect of the Earth's rotation is negligible, as its velocity is approximately 1/100 that of its orbital velocity. The situation is shown in Figure 7.5, where Figure 7.5(a) is an hypothetical stationary Earth. Here the actual light path from a star propagating vertically (φ_p=90°), in the propagation medium, is indicated by the dotted line, i.e. the path that would be seen with chalk dust and a laser. For the orbiting case, illustrated in Figure 7.5(b), the Earth and telescope are moving relative to the stationary star and medium, here the motion is considered perpendicular to the light path. The classical time in the stationary frame τ_p for the light to travel through a vertical height W is W/c. In this time the telescope has moved a distance a_o'= $o\tau_p$. Thus the classical observer angle φ_o' for light to pass down the telescope is:

$$\tau_p = W/c, \quad a_o' = o\tau_p = Wo/c, \quad \tan\varphi_o' = W/a_o' = c/o = 1/M \qquad (7.49)$$

Including the very small relativistic modifying effect, the vertical velocity c in the y direction is transformed, between the stationary medium and the moving frame, becoming αc, as shown in Figure 7.2. Thus the complete classical plus relativistic effect becomes:

$$\tau_o = W/\alpha c, \quad a_o = o\tau_o = Wo/\alpha c = a_o'/\alpha, \quad \tan\varphi_o = W/a_o = \alpha c/o = \alpha/M \tag{7.50}$$

$$(\theta_o)_{abr} = 90^o - \varphi_o$$

For most purposes α can be taken as unity and $\varphi_o = \varphi_o'$. $(\theta_o)_{abr}$ is the angle in the moving observer frame, i.e. the angle the telescope has to be tilted in front of the star to be viewed, in the direction of motion. Physically, the effect is analogous to an open truck, carrying an inclined tube, moving through a tunnel with water dripping vertically from the roof (stationary frame). On board the truck (moving frame), the drip paths will be seen to be at an observer angle θ_o, tilted vertical in the direction of motion. This is the angle the tube would have to be tilted for the drips to pass through the tube. The angle of course is independent of the length W, it is a resolved angle in the direction of motion, the water still falls vertically. This was essentially Bradley's argument; except that he used the analogous example of a wind aberration angle on the flag's direction, on his sailing boat, with forward speed.

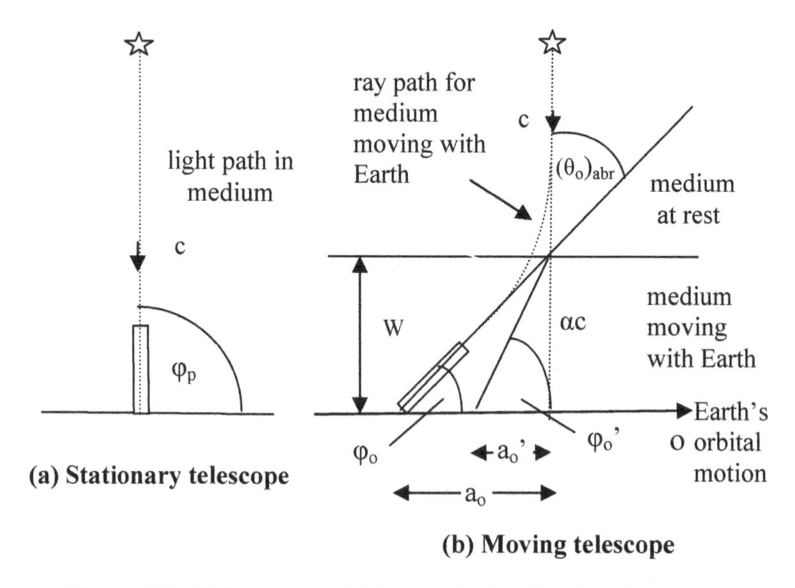

Figure 7.5 Telescope orbiting with the Earth relative to a stationary propagation medium and propagating starlight

It was believed that there was no propagation medium, as the medium would affect the aberration angle. This is not the case, the medium orbiting with the Earth does not affect the aberration angle. In fact the medium is required to provide a smooth transition for the light from the stationary to the moving medium as shown in Figure 7.5(b). The angle becomes an actual angle in the medium, not a resolved angle between the Earth's motion and the speed of light, which would have happened without the medium. W is now the extent of the medium moving with the Earth, considered in Section 3.1, Chapter VIII.

Thus the vertical ray angle φ_p in the stationary propagation medium becomes the observer aberration angle φ_o in the moving frame. This same angle is carried down into the telescope. There is no change in angle here because there is no relative motion between the telescope and the medium, and therefore no Fresnel partial convection can take place. Also there is zero angle of incidence at the telescope lens, giving no Snell refraction. This explains why filling the telescope with water or any other dense material has no effect on stellar aberration. Without the medium moving with the Earth these effects cannot be explained, thus again confirming the medium's presence

Formally, using the relativistic addition of velocities, light travelling vertically in the stationary propagation medium, in Figure 7.5(a), is defined by u_{py}=c and u_{px}=0. In the x direction in the moving frame, from equation 7.26, $u_{ox}=(u_{px}+o)/(1+u_{px}M/c)=o$. In the y direction in the moving frame, from equation 7.30, $u_{oy}=u_{py}\alpha/(1+u_{px}M/c)=c\alpha$. Thus, transforming from the stationary propagation medium to the moving frame one has:

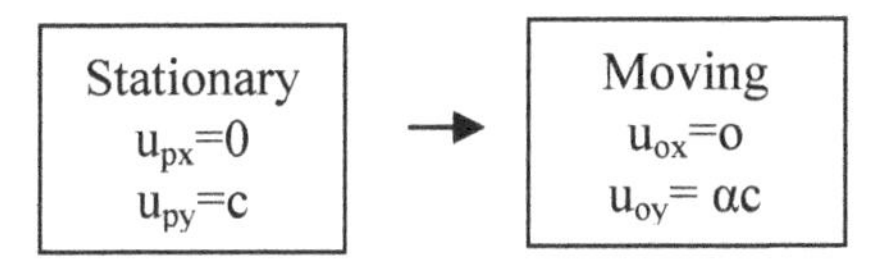

The propagation angles in the stationary and moving frames become:

$$\tan \varphi_p = u_{py}/u_{px} = c/0 = \infty, \quad \varphi_p = 90^\circ$$

$$\tan \varphi_o = u_{oy}/u_{ox} = c\alpha/o = \alpha/M \qquad (7.51)$$

For M=0.6, α=0.8, the telescope's elevation angle in the moving frame is φ_o=53°, giving the Bradley observer angle θ_o =90°-φ_o=37°. For the classical effect alone α=1, the corresponding angles are φ_o'=59° and θ_o =31°. Thus the second order relativistic (Lorentzian) effect is minor compared to the first order classical (Galilean) effect. Thus one can neglect the relativistic effect at Earth speeds ($M=10^{-6}$), but the classical effect is lost if the propagation medium is removed. The dotted lines in Figures 7.3, 7.4 and 7.5 are the actual light paths through the stationary medium, i.e. those paths that can be seen with chalk dust and a laser beam that relativity cannot account for.

4.4 Light convection and medium motion

Light convection is the partial increase in light speed in the direction of motion of a dense medium moving relative to its light source and vacuum medium. Unlike classical media, 100% light convection is not possible, because the light is already propagating in the vacuum. Fresnel (1818) developed and Fizeau (1851) confirmed the convection formula. It is formally proved using the propagation medium and the relativistic addition of velocities. Let $u_{px}=c_p$ be the propagation velocity in the stationary dense medium and $u_{ox}=c_o$ be the resulting velocity in the moving frame moving with velocity o with respect to the stationary medium, as illustrated in Figure 7.6. Here the convective speed c_o relative to the stationary medium speed c_p, is from equation 7.26:

$$c_o=u_{ox}=(u_{px}+o)/(1+u_{px}M/c)=(c_p+o)/(1+c_po/c^2) \qquad (7.52)$$

Using $(1+x)^{-m}\approx 1-mx$ gives

$$c_o \approx (c_p+o)(1-c_po/c^2) \approx c_p + o - c_p^2o/c^2 - c_po^2/c^2 \qquad (7.53)$$

Neglecting the last term, as $c_p^2 >> c_p\, o$

$$c_o \approx c_p + o - c_p^2 o/c^2 = c_p + o[(1-n^{-2}] \quad \text{where} \quad n = c/c_p \qquad (7.54)$$

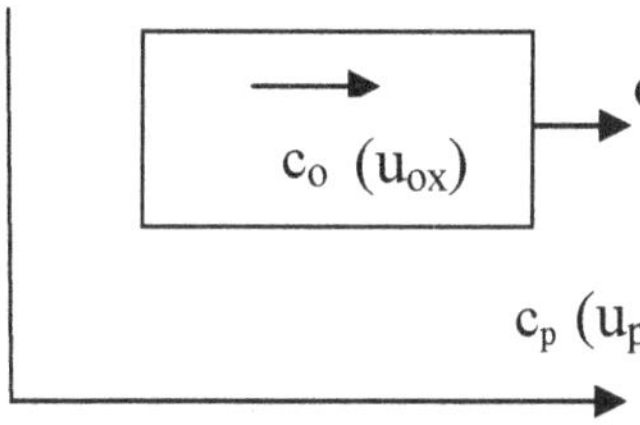

o

$c_o\ (u_{ox})$

$c_p\ (u_{px})$

Figure 7.6 Light convection by a dense (n>1) moving medium

Equation 7.54 is Fresnel's convection formula, where n is the refractive index, i.e. the ratio of the propagation speed of light in a vacuum compared to the more dense medium. For example if n=1.4, $(1-n^{-2}) = (1-1.4^{-2}) \approx 0.5$ i.e. 50% convection velocity of 'o'. In terms of the medium electrical parameters, referenced to those in a vacuum, the refractive index is given by:

$$n = (\mu_r\, ę_r)^{1/2} \qquad (7.55)$$

where μ_r is the medium's relative permeability (electrical inertia) and $ę_r$ is the relative permittivity, (electrical stiffness). For most materials, μ_r is very close to 1 at optical frequencies, therefore n is approximately $(ę_r)^{1/2}$. Thus under the influence of gravity, where the medium is compressed, $ę_r$ changes, μ_r remaining basically unaffected.

5 Gravity and Space-Time

Electromagnetic waves propagate at the speed of light 'c', irrespective of source or observer motion 'v', with respect to the propagation medium. Velocities add relativistically with respect to the medium. Time and space contract with motion relative to the propagation medium. This allows systems and space travellers, moving relative to the medium, to age less than their stay-at-home counterparts. Similarly, mass m* increases, through kinematic motion relative to the medium, by approximately:

$$\alpha^{-1} = (1-(v/c)^2)^{-1/2} \approx 1+1/2(v/c)^2 , \quad v/c<<1 \quad (7.56)$$

Thus

$$m^*\alpha^{-1} = m^*[1+(½)(v/c)^2]=m^*+m', \quad m'= (½)m^*(v/c)^2 = E/c^2 \quad (7.57)$$

$$\text{where } E = (½)m^*v^2 \text{ giving } E = m'c^2 \quad (7.58)$$

Thus m' is the effective mass increase through motion.

Therefore, those parts of Einstein's relativity based on the LT, including invariance of the speed of light, time and space contraction, transverse Doppler, particle lifetime, relativistic mass and energy, and his famous equation between mass and energy $E=m'c^2$, are all supported by this revised theory based on a stationary propagation medium. Also, the revised theory does not affect the two general equations of GR. For example, in Einstein's GR (1915), his equation of motion with respect to absolute space can be written as:

$$R_{\mu v} - 1/2 \, R \, g_{\mu v} = 8\pi \, G \, T_{\mu v} \qquad (7.59)$$

The left-hand side of the equation describes the distortion (compression) of the propagation medium compared to its undistorted free space value. The right-hand side defines the energy and momentum causing the distortion. $R_{\mu v}$ and R are Riemann (1866) tensor and scalar quantities, which specifies the distortion, $g_{\mu v}$ is the generalization of the Minkowski metric (1908), G is Newton's gravitation constant and $T_{\mu v}$ is the energy momentum tensor, which specifies the flux of the μth component in the vth direction. Basically, the equation determines how energy momentum distorts the propagation compared to the absolute free space value. The other important equation in Einstein's GR is the space-time four vector. The source time t_s and space x_s, in the moving source frame, expressed in terms of the stationary propagation medium time t_p and space x_p, y_p and z_p, can be written as:

$$x_s = (x_p^2+y_p^2+z_p^2-c^2t_p^2)^{1/2} \qquad (7.60)$$

Considering motion in say the x direction and dividing by c

$$t_s = [(x_p/c)^2-t_p^2]^{1/2} \qquad (7.61)$$

Equation 7.61 represents a right angled triangle of vertical axis t_p and horizontal axis x_p representing the stationary propagation medium, where the hypotenuse is $t_s=x_s/c$ represents propagation in the moving frame. It can be space like (large distances in small time) with a gradient close to 45°, or it can be time like (small distances in large time) close to the vertical axis (90°), depending on the speed with which the system moves through the stationary propagation medium. Rearranging equation 7.61 gives:

$$t_s= [(x_p/c)^2-t_p^2]^{1/2}= j[t_p^2-(x_p/c)^2]^{1/2} = j\, t_p\, [1-(x_p/ t_p\, c)^2]^{1/2}$$

$$= j\, t_p\, [1-(s/c)^2]^{1/2}= j\, t_p\, \alpha_s \qquad (7.62)$$

where

$$j = (-1)^{1/2}, \quad \alpha_s= [1-(M_s)^2]^{1/2}, \quad M_s = s/c, \quad s= x_p/ t_p$$

$$t_p = x_p/c, \quad t_s = x_s/c, \qquad (7.63)$$

If t_p is in years, x_p will be in light years (LY). For example if M=0.6, $\alpha = 0.8$, $x_p = 3$LY, $x_s = \alpha_s x_p = 0.8x3 = 2.4$LY, $t_p = x_p/M = 3/0.6 = 5$ years, $t_s = j\alpha t_p = j0.8x5 = j4$ years, which is a perfect imaginary right-angled triangle with sides 3, 4 and 5. However, the largest side is the opposite (t_p) not the hypotenuse (t_s). Time t_s in the moving frame is smaller (slower), by α, and is imaginary compared with time t_p in the stationary propagation medium. For speeds greater than the speed of light $\alpha \rightarrow j\alpha$, making $t_s = -t_p\alpha$, i.e. time in the moving frame is now real again, and has the potential to run backwards (see Figure 6.2, Chapter VI for further information).

Equations 7.59 and 7.60 in combination are apparently difficult to solve generally. However, a propagation medium is essential for their solution. Schwarzschild's (1916) metric using spherical geometry is obtained using a propagation medium devoid of matter (vacuum), apart from the source matter situated at the centre of the coordinate system. Again, using a frame at rest in absolute space (stationary propagation medium) and assuming the medium to be homogeneous and isotropic, the Robertson-Walker (1920) metric in cosmology is obtained. However, a simplified Schwarzschild's model is used in Sections 6 and 7 of Chapter V to demonstrate the basic effect of gravity on the propagation medium (time and space compression). These are compared with Lorentz's time and space contraction of system moving through the medium.

6 Conclusions

The Lorentz Transform which is based on observer motion with respect to a stationary propagation medium is extended graphically for both source and observer motion relative to the propagation medium. It is shown that relativistic addition of velocities is based on Lorentz transform and its propagation medium. The actual propagation paths through the medium are identified for some important applications. Reflection from a railway carriage window, Bradley's stellar aberration and convection of light by a moving dense medium. It is shown in these applications that the light propagation is via the propagation medium, not through an unknown ether-less mechanism.

Chapter VIII:

Situation Around Planets

1 Introduction

In this chapter the nature of gravity is considered and the medium around planets investigated. Gravity is considered to be a residual difference electric field between distributed dissimilar charges within atoms and molecules. Gravitational mass has an attractiveness similar to the electric charge, but 10^{-20} weaker. Also the residual difference electric field from the total gravitational mass in the universe is considered to create inertia and provide a universal omni-directional reference field. The medium is attracted and compressed by gravity. It surrounds and orbits planets within the planets Gravitational Field of Dominance (GFOD), in the presence of its sun's gravity.

The medium orbits with planets in the near field and is at rest in space in the far field. If the planet rotates it can form an Electro-gravitational Boundary Layer (EGBL) between the 'stationary' surrounding medium and the rotating planet's surface. The difference between source and observer motion angles are also discussed. The effect of Global Positioning Systems (GPS) and satellite communication signals passing through the 'stationary' GFOD medium surrounding the Earth is considered. Measurements with respect to the medium on the planet and out in space away from gravitational bodies are suggested.

2 Nature of Gravity

Gravity has similar properties to the electric field. Both are attractive, their intensity is dependent on the inverse square law with distance, they use the same medium and have a propagation (retarded) time delay. Generally, a distribution of positive and negative charges will produce internally an attractive force and externally a residual difference attractive electric field. This residual field will tend to zero as the distribution size goes to zero. But for a finite distribution this field remains finite, capable of attracting similar multipole charge distributions contained in matter. This appears to be the basis of gravity explaining how dissimilar charges from gravitational matter always attract. Thus unsteady electric (electromagnetic), steady difference electric (gravitational), and residual gravity fields throughout the universe (inertial) all appear to be electrical, using the same propagation medium. This allows the Lorentz transform, gravity and accelerating frames to be compared directly.

2.1 Atomic modelling

Building on the model of discrete quanta Planck (1900), the Bohr atom (1913) and then the wave representation of particles by De Broglie (1924), this new century saw the extraordinary development of quantum field theories. Here strange happenings were revealed approaching the plank scale, including quantum field jitters, characterized by the Heisenberg's Uncertainty Principle (1927). Quantum Electro Dynamics (QED) were then developed by Dirac (1928), from extending Schrödinger's Equation (1927) and including the relativistics of the Lorentz transform. However, trying to unite the large scales of gravity (GR) with the small scales of QED, using 'point' atomic models (PAM), causes singularities (infinities) in their field equations. It appears that this impasse can be avoided, quite naturally, by replacing the compact models with non-compact finite structures.

2.2 Strings

One approach, which has had some success, that avoids singularities, is to represent the point model with finite string structures. This appears to have the potential to represent the basic constituents of the universe, mathematically, without a physical model. The implementation is to represent space by the usual three dimensions, plus 6 or 7 curled up 'dimensions' that can be fine tuned and 'plucked' to resonate in various vibration modes. The extra 'dimensions' appear to behave more like complex 3D structures, rather than extra dimensions. These structures could perhaps, with sufficient detail, represent any particle or force in nature.

String theory has been investigated, seriously, now for over 20 years. The main problem seems to be to construct and solve, accurately, the complex mathematical models of 9 or 10 features of string geometries that can represent nature's particles and forces. Kaluza (1921), just after the First World War, seemed to be on the right track. He approached Einstein with an extra dimensional model; three representing regular space, one time and one equivalent to EM force, Einstein after some consideration decided he was not impressed.

2.3 Distributed atomic structures

After the big bang, the universe is considered to have expanded rapidly (inflation) through a repulsive force. At these initially high densities and temperatures the electric fields of the prototype protons and electrons would have been non-existent. After sufficient expansion and cooling, the matter finally condensed into regular atoms of electrons, protons and neutrons, with the familiar force of attraction between them (gravity). These atoms, the basic building blocks of matter, have complex structures. Rather than representing by strings, a more physical and realistic approach is to consider the actual finite dimensions of these Distributed Atomic Structures (DAS). According to the Standard Model these structures are considered to contain 16 elementary subatomic particles, 12 fermions which have mass, and 4 bosons which are force carriers, plus a second set of antiparticle equivalents.

In simple terms, the fermions have 6 varieties of quarks (up-down, strong-charm and bottom-top), and 6 flavours of leptons (electron, muon, tau, and neutrino versions of the electron, muon, tau). The bosons contain: (i) The strong steady nuclear force (gluons), which bind the quarks together in the nucleus. (ii) The weak nuclear forces (W^+, W^- and Z) which are concerned with nuclear emission, absorption and decay. (iii) The electromagnetic force (varying E field through charge variations), which is responsible for light emission (photons) and (iv) gravitational force variations, which are thought to create gravity waves (gravitons) caused by gravitational matter variations. Further composite particles, having various charge values (plus and minus integer and zero), also exist.

2.4 Electrostatic attraction

Fortunately, most of the matter in the universe is thought to be comprised of just negatively charged electrons and two types of charged quarks, having fractional values of charge q, an up quark 'u' (q=2/3) and a down quark 'd' (q=-1/3). The main field contributors are therefore the proton q=u+u+d=2/3+2/3-1/3=1 and the electron q=-1. Even possibly the neutron q=d+d+u=-1/3-1/3+2/3=0, if its quarks, under high pressure, can be reconstructed into positive and negative charges. Using a point atomic model an equal number of negative electrons and positive protons, with their corresponding electric fields cancel, creating a neutral atom. However, a finite distribution of dissimilar charges will always produce an attractive steady residual difference field. Although diagonal charges repel in the multipoles, adjacent charges attract and are closer together, they therefore dominate. This can be easily demonstrated with dissimilar magnetic or charged poles, as illustrated in Figure 8.1. This simple basic attractive mechanism appears to play an important part in the attraction of atoms and molecules (protons and electrons) and therefore gravity.

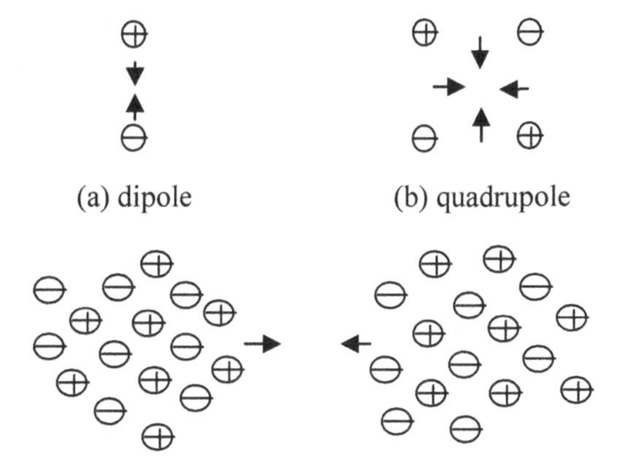

Figure 8.1 A finite distribution of dissimilar charges will always produce a residual attractive field forming the basis of gravity

2.5 Van der Wall's attractive dipoles

According to van der Waals (1873), dipoles from molecules with permanent bipolar fields and induced dipoles, causes an attraction between individual atoms and molecules. These attractive dipole fields, in the near field, determine their liquid and gaseous phases and the pulling together of mercury beads and soap films. They are local, having a very rapidly reducing field strength with distance. Also London (1937), predicted that electrons orbiting at large distances around atomic nuclei repel (displace) electron orbits of neighbouring atoms and molecules. These unbalanced orbits then create a mass-spring induced dipole field, oscillating at imaginary non propagating frequencies, resulting in a rapidly decaying reactive field with distance. However, low atomic number molecules can have high and low energy levels. These result in strong and weak attractive near fields creating both solids and gases with similar atomic numbers, i.e. their attraction is not proportional to the atomic number.

2.6 Gravitational field

Although these van der Waals type attractive fields are electrostatic, they do not appear to be gravitational, they are local fields holding the molecules together. For gravity we are looking for attractive electrical dipole fields that propagate into the radiation far field, i.e. decaying according to the inverse square law and whose strength (weight) tends to be proportional to the atomic number. The model that best fits this description is the Atomic Residual Difference Field (ARDF) whose attractive field is proportional to the atomic number (number of pairs of dissimilar charges (dipoles) in the atom from protons and electrons). From quantum mechanical grounds, the difference residual field should form in the reactive near field, at the atomic level.

The field summation from all atoms and molecules in the vicinity would then be attractive and decay into the propagation far field according to the inverse square law, the same as gravity. Equating the electric field E, given by $E=kq^*r'/r^2$ where $k=8.99x10^9\ Nm^2/C^2$ and q^* is an attractive dissimilar charge in Coulombs. And the gravity field g, given by $g=Gm^*r'/r^2$ where $G=6.67x10^{-11}Nm^2/kg^2$, and the attractive mass m^* is in kg. Also r' is the unit vector along the distance radius r. Therefore gravitational mass m^* has an equivalent charge q^*given by:

$$q^*=(G/k)m^*=(6.67x10^{-11}/8.99x10^9)m^*=0.74x10^{-20}m^* \qquad (8.1)$$

which creates a potentially very weak attractive field. Other equivalents are then:

$$g\equiv E,\ \text{ force } F=m^*g\equiv q^*E,\ \text{ potential } v=\int gdr\equiv\int Edr,$$

$$\text{energy } U=Fdr=\int m^*gdr=m^*v\equiv\int q^*Edr=q^*v \qquad (8.2)$$

2.7 Inertial field

Whatever the actual details of the ARDF, it can be considered to be finite from the almost limitless gravitational matter, seen and unseen throughout the universe. Although this field reduces with the inverse square of the distance from its gravitational source, the field is enhanced at any point in the universe, from all known and inferred gravitational matter of equivalent volume charge density ρ* lying on an expanding sphere whose incremental annulus ($4\pi r^2.dr$) increases as the square of the distance, compensating the inverse square loss. The modulus of the total field (inertial field) contributing to any point in space then becomes:

$$|E| = k \sum |q^*| / r^2 \quad \text{where} \quad \sum |q^*| = \int (\rho^* 4\pi r^2)\, dr,$$

$$|E| = \int (k\rho^* 4\pi)\, dr = k\, \rho^* 4\pi r, \quad \text{where} \quad r = \infty \qquad (8.3)$$

Only through symmetry does the ARDF tend to zero, E=0. but the modulus of the inertial field $|E|$ is limitless (large). Thus from the total matter in the universe, the ARDF provides a non zero Universal Gravitational Reference Field (UGRF) over all space. This scalar potential energy field is available to resist motional change of mass (energy) in motion, creating mass inertia. The $|E|$ field is also a measure of the magnitude of the attractive 'dark' energy, resisting, (slowing down) the gravitational mass expansion in the universe.

The medium could then provide the complementary repulsive 'dark' energy, overcoming the global gravitational attraction, through the repulsive expansion of the medium. Unlike unsteady EM and gravitational waves, the UGRF, from steady charges throughout the universe behaves as a steady field. The ARDF, being a difference E field, has no mass and spin 1. Whereas, the UGRF, being a resultant scalar potential field from difference E fields from all directions across the universe, has zero intensity (force), no mass and no spin, making it difficult to detect, but it has finite potential energy. This field resembles the Higgs field (1964).

The medium that supports this field, is discussed in Section 2, Chapter III. It has a non-stressed (free space) electrical medium inertia (permeability), $\mu=1.25\text{x}10^{-6}$ N/A^2 and an electrical rigidity term (permittivity), ę=8.85x10^{-12} F/m. The medium has no viscous terms to impede the steady movement of inertial mass through it. But its field does possess potential energy that opposes changes in motion (kinetic energy). If the medium is gravitationally compressed locally, for example around a heavy body, medium contours of reducing compression will result. The path of a light beam (light bending) will then depend on its initial direction and on the gradient of the medium contours surrounding the body.

2.8 Inertial and gravitational mass

As Einstein kept reminding us, all forms of energy (potential and kinetic) are equivalent to inertial mass. It appears that inertial mass of subatomic particles, complete atoms and molecules, results through the binding energy of the system, the energy required to create the structure. For example, in the proton only 2% of its binding mass is from its individual components, the other 98% is accountable through the strong binding force (energy) of the gluons holding the proton together. Inertia is the resistance to accelerating binding mass. As in electrons in orbit around an atom and J J Thompson's e/m experiment (1897), where gravity has no part, only inertial mass (no g or G constants involved).

It appears that it is not the inertial or binding mass m_b that causes gravity. It is a quantity of energy, having no substance or solidity in itself and no attractive capability. Attraction of matter (gravity), appears to be the Atomic Residual Difference Fields (ARDF) between finite distribution of dissimilar charges within gravitational matter (atoms and molecules) having a total gravitational mass m_g. Here m_b=m_g, but m_b can exist without m_g, as in the electron having mass but no mass attraction. Although neutron stars are nominally neutral, it appears, under the huge compacting pressures, their gravitational matter, including neutrons (which are comprised of positive and negative fractional charges) must undergo some kind of charge reconstitution of their atomic quarks into positive and negative free charges (currents), creating their intense magnetic and gravitational fields.

Thus: i) Light is a varying E field from varying charges. ii) The ARDF is the residual steady difference E field from primarily proton and electron charge dipole distributions. iii) Gravity is generated from the ARDF from a large number of atoms and molecules. iv) The UGRF is the residual field from the ARDF from all directions in the universe, from the total gravitational matter in the universe. v) The UGRF propagating in the medium at rest in space provides the inertial field and absolute reference. All these effects have a common basis i.e. they are all electrical, they all depend on the electric charge.

Without charge there could be no electric field, no ARDF (gravity) and no UGRF (inertial field). Without varying charge there could be no varying electric field (light). Newton (1687) claimed that accelerative motion was with respect to absolute space. Mach (1887), claimed that it was with respect to the distant matter of the universe. These concepts support the UGRF and propagating medium. All forms of motion, including acceleration and constant motion, can be detected through observed motion with respect to the propagation medium at rest in space.

3 Medium Around Planets

3.1 Gravitational Field of Dominance (GFOD)

It has been shown in Section 6, Chapter II, that there is considerable evidence for both the propagation medium at rest in space and one that moves with gravitational bodies. The ARDF from local gravitational mass concentrations dominates the propagation medium closest to the body. It is then attracted, compressed and moves with the body. If the body rotates, the medium in contact with its surface can rotate. The existence of the medium moving with the Earth's surface is supported by the MMX (1887), Section 2, Chapter I and Sagnac (1913), Section 5.1, Chapter IV.

The 'stationary' medium surrounding and orbiting with the Earth is confirmed through the existence of the medium boundary layer between the rotating Earth and the surrounding stationary medium, established by M&G (1925), Section 5.3 and by H&K (1972), Section 5.7, Chapter IV. If there was no 'stationary' medium surrounding the Earth, then propagation would have been very asymmetrical around the Earth, through its motion relative to the medium at rest in 'space', which has not been measured.

At large distances from gravitational matter, the local attraction becomes weak reverting back to the inertial UGRF pervading through the propagation medium at rest in space, as supported by Bradley (1725), Section 4.3, Chapter VII; De Sitter (1913), Section 3.3, Chapter VIII, Brecher (1977), Section 4.4, Chapter IV and COBE (1992), Section 6, Chapter II. Thus there is a Gravitational Field of Dominance (GFOD) where the medium, is attracted, surrounds and moves with gravitational bodies. The extent of its dominance appears to be controlled by the body's individual gravitational field compared to other gravitational fields in the vicinity, as illustrated in Figure 8.2.

From equations 5.33 and 5.34, Chapter V, the gravitational time slowing $\Delta\tau_3$, at a distance of R_{u3} away form an isolated gravitational mass m_3, compared to the UGRF time τ_u is given by:

$$\Delta\tau_3 \approx -M_g/2\Delta\tau_u, \quad M_g = 2Gm^*_3/c^2R_{u3} \quad (8.4)$$

Where $G=6.67\text{x}10^{-11}$, for a small gravitational mass m^*_1 in the presence of a large gravitational mass m^*_2, separated by a distance $D_{1,2}$, the Schwarzschild's equal time radius $R_{S,1,2}$ around the small mass, based on equal time slowing, $\Delta\tau_1 = \Delta\tau_2$, then becomes:

$$Gm^*_1/c^2 R_{S,1,2} = Gm^*_2/c^2(D_{12}-R_{S,1,2})$$

Assuming $m_2 >> m_1$, then $D_{1,2} >> R_{N,1,2}$,

$$R_{S,1,2} \approx (m^*_1/m^*_2)D_{1,2} \quad (8.5)$$

Or in terms of equal gravitation $g_1 = g_2$, becomes:

$$G\, m^*_1/(R_{N,1,2})^2 = G\, m^*_2/(D_{12}-R_{N,1,2})^2$$

$$R_{N,1,2} \approx (m^*_1/m^*_2)^{1/2}D_{1,2} \quad (8.6)$$

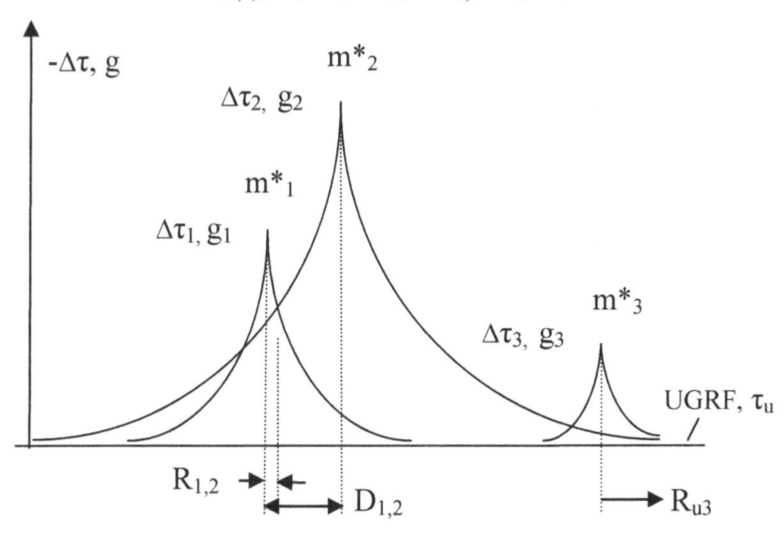

Figure 8.2 Gravitational Field of Dominance (GFOD) of gravitational masses m^*_1, m^*_2, and m^*_3, protruding through the sea of the Universal Gravitational Reference Field (UGRF)

The extent of the Earth's GFOD, in the presence of the Sun, is predicted from equation 8.5 to be $R_S \approx (6x10^{27}/2x10^{33})x(1.5x10^{11}) =4.5x10^5$m, and using equation 8.6, $R_N \approx (6x10^{27}/2x10^{33})^{1/2}x(1.5x10^{11}) =3.2x10^8$m. The ratio of the Earth's GFOD compared to its radius is $4.5.10^5/6.10^6 \approx 1/10$ and $3.2x10^8/6x10^6 \approx 50$, using equations 8.5 and 8.6 respectively. Schwarzschild (1916) compressive inverse law produces only a small distance (1/10 of the Earth's radius). Possibly, Schwarzschild compression contributes to the EGBL effect, discussed further in Section 4.3. Perhaps the effect here is dominated more by the proximity of the surface of a large amount of matter rather than the complete body. Whereas, Newton's attractive inverse square law field appears to cause the GFOD effect in the far field.

The extent of the Earth's GFOD, compared to its orbital radius around the Sun is $3.2x10^8/1.5x10^{11}=2.2x10^{-3} \approx 1/500$, using equation 8.6. Other Solar System GFOD's, in the local group, using equation 8.6, are Mercury=$0.2x10^8$, Venus=$1.5x10^8$ and Mars=$1.3x10^8$. In the distant group: Jupiter=$2.4x10^{10}$, Saturn=$2.3x10^{10}$, Uranus=$6x10^{10}$, Neptune=$3.2x10^{10}$ and Pluto=$1.3x10^{10}$. If the gravitational matter is in motion, it appears that its GFOD with its surrounding medium moves with it. Therefore, the GFOD, through its gravitational dominance, appears to control the propagation medium motion around planets. Whereas, mechanical system dynamics on planets, for example Foucault's pendulum, appears to takes its reference from the omnipresent UGRF. Accurate measurements of the extent and gradient details of the GFOD for different sized planets, in the presence of more massive bodies, would be useful.

3.2 Gravitational and motional time

Lorentzian (relativistic) time slowing with motion with respect to the propagation medium, from Section 7.1, Chapter I, and in detail in Section 4, Chapter V is given by $\Delta\tau_s \approx \Delta\alpha\Delta\tau_p \approx -\Delta\tau_p M^2/2$. Repeating the calculation, the Earth rotating with surface velocity at the equator of

460m/s, $M=460/3x10^8=1.5x10^{-6}$ giving $\Delta\alpha\approx M^2/2=1.5^2x10^{-12}/2=1.1x$ 10^{-12}. The time slowing τ_{rot}, through rotation relative to the planet's GFOD time, for each 24 hour period ($8.6x10^4$s), is therefore $\tau_{rot}\approx$-$8.6x10^4x1.1x10^{-12}\approx$-95ns per day. For the Earth's orbital velocity around the Sun at about 30km/s ($M\approx0.0001$), its time slowing compared to the Sun's time τ_{sun} is $\Delta\tau_{orb}\approx-\tau_{sun}M^2/2\approx-8.6x10^4x10^{-8}/2\approx$-0.4ms/day. For the Sun and Solar System's time slowing travelling through space at about 300 km/s (M=0.001), their time slowing $\Delta\tau_{sun}$ compared to the universal reference (UGRF) time τ_u is $\Delta\tau_{sun}\approx-\tau_u M^2/2\approx$-$8.6x10^4x10^{-6}/2\approx$-40ms/day ($\approx$100x that of the Earth orbiting around the Sun). Finally, the relativistic time τ_v, for motion on Earth, for example Picasso et al (1977) rotating Muon's time, Section 5.6, Chapter IV is 30 times slower through its velocity 'v' relative to the planet's rotating surface time τ_{rot}. However, most velocities on Earth are much less than the Earth's rotational speed, thus:

$$\tau_v < \tau_{rot} < \tau_{orb} < \tau_{sun} < \tau_u. \qquad (8.7)$$

$$\tau_v < 95\text{ns} < 0.4\text{ms} < 40\text{ms} < \tau_u$$

If $\Delta\tau_s$ is the source relativistic time change through motion and $\Delta\tau_g$ is the gravitational time change in the vicinity of the source, then from Section 7, Chapter V, the total source time slowing, from equations 5.30, 534 and 5.39, is approximately :

$$\Delta\tau_T \approx \Delta\tau_s + \Delta\tau_g \qquad (8.8)$$

where $\Delta\tau_s$ and $\Delta\tau_g$, are both negative (time slowing). It appears the gravitational time slowing on the Earth's surface, from its own gravitational mass and through the Sun's gravitational mass, is approximately -0.06ms/day and -0.7ms/day, respectively, the distant Sun has the bigger influence.

3.3 Medium within GFOD

There is no propagation medium in the conventional (mechanical) sense. However, there is a well defined electrical medium of electrical inertia μ and stiffness ę, as discussed in Section 2.4, Chapter IV. This medium is energized by steady electric fields, difference electric fields and propagates unsteady electric fields. The medium can be at rest in space or move with gravitational bodies. The 100% light propagation with the medium and its light source, moving with gravitational bodies, should not be confused with light convection, where the medium moves relative to the source. Through unity refractive index there can be no light convection. But the medium being an electrical compressible fluid can refract light through medium compressions by gravity or though medium velocity gradients.

Light emanates from an EM source and propagates at the speed of light from the source with respect to the medium within the GFOD. Moving with the body, as demonstrated by: (i) Hafele and Keating (1972), Section 5.7, Chapter IV, in their high flying aircraft's symmetrical time slowing relative to the stationary GFOD medium. (ii) Saburi et al (1976), Section 3.7, where communication satellite motion relative to the GFOD produces a PTA of 300ns. and (iii) GPS (1992), satellite motion relative to the GDOD, Section 3.5, produces a PTA resulting in positional displacement of 30 m on Earth. These measurements are all predicted through classical motion relative to the stationary GFOD medium orbiting with the Earth.

The EM wave transmission and reception process on Earth is not caused through dragging of light, nor through Fresnel's (1818) partial light convection. The EM source and its propagation medium, within its GFOD, simply move together with the heavy gravitational body. For moderate sized bodies, small event horizons, the light will leave the body and slow down as it adjusts to the propagation speed in the stationary medium. Thus, the propagation speed locally in the medium moving with the body, plus the body speed, can exceed the propagation speed in free space. Of course if there is a very intense gravitational field, for example, close to a black hole, the light and its medium will be severely compressed and will be attracted strongly, spiralling towards the hole.

De Sitter (1913) showed that light from high speed binary stars is Doppler shifted normally, at large distances from the stars, similar to sources moving relative to the medium at rest in space. This is in spite of very massive neutron stars attracting the medium within their GFOD. The light is not distorted through wave overlap or gaps that would have resulted if the medium, which moves with the stars in their elliptical orbits, had moved out to the far field. Ritz (1908) proposed an emission theory, together with the extinction length. This basically occurs when the speed of light adds to the source speed, close to the source, and then after a certain distance returns to the speed in the stationary medium. The situation is considered further by Fox (1965) and by Brecher (1977), discussed in Section 4.4, Chapter IV.

The process is illustrated in Figure 8.3. Figure (a) shows concentric propagation circles expanding at the speed of light, close to a large gravitational source and its GFOD. Figure (b) shows the situation in the far field for a small moving source. Here non-concentric circles are now seen whose centres are equally spaced in the stationary medium in the far field. According to Huygens' Principle (1690), each wave front moves normal to its original source emission position, producing a new wave front with an angle $(\theta_s)_{orb}$. Thus a light ray propagating normally (90°) to the direction of the star orbital motion, moves away from the source at this source motion angle, behind the normal, given by:

$$(\theta_s)_{orb} \approx \tan^{-1}(M_{orb}), \quad M_{orb}=s/c \quad (8.9)$$

Here 's' is the orbital velocity and 'c' is the light speed in the medium. The Earth's orbital velocity is 30 km/s ($M \approx 3x10^4x3^{-1}x10^{-8} = 10^{-4}$)

giving a source orbital ray angle from equation 8.9 of $(\theta_s)_{orb} \approx \tan^{-1}(M) = 5.7X10^{-3}$degrees.

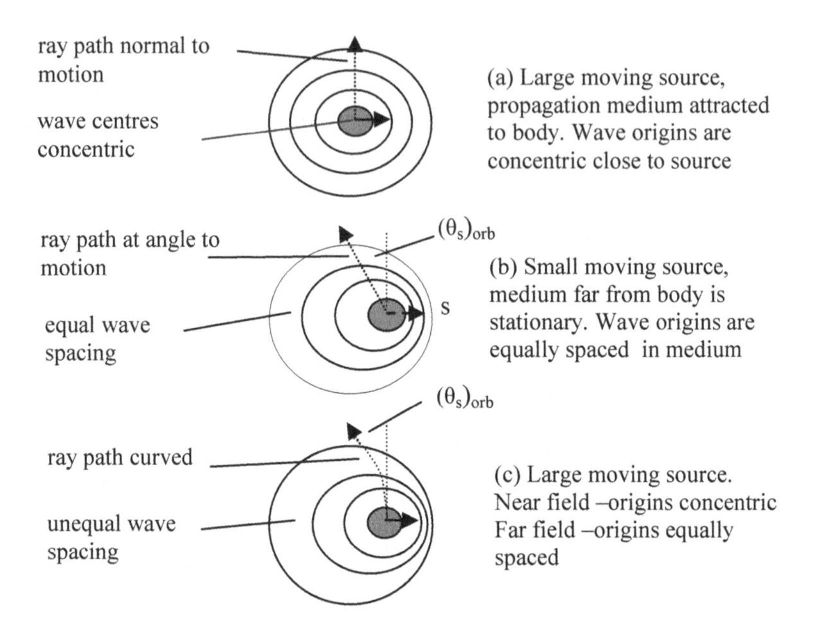

Figure 8.3 (a) EM medium moving with a large source within its GFOD (near field), (b) small source moving relative to the medium at rest in space (far field), (c) a combination of (a) and (b)

This source orbital angle $(\theta_s)_{orb}$ is the complement to the observer motional angle $(\theta_o)_{abr}$, in Bradley's stellar aberration, which is forward to the normal in the direction of motion, as discussed in Section 4.3, Figure 7.5(b), Chapter VII. Here a light ray going from the stationary source and medium in space to a moving planet and observer (telescope) appears at an angle $(\theta_s)_{abr}$ in front of this source to be viewed. In Figure 8.3 (c), the circles are shown as concentric circles in the medium close to and moving with a large gravitational body's GFOD, similar to Figure 8.3 (a). At large distances from the body the wave centres will again appear equally spaced moving in the stationary medium, little affected by the local medium moving with the body. Thus the body's GFOD acts as a buffer as light rays become curved as they pass from the moving source and medium to the 'stationary' medium.

For small/moderate gravitational bodies, the effect of local medium moving with the body, will appear transparent in the far field. The dotted lines in Figure 8.3 are actual light paths through the medium. Thus source and observer motion, with respect to the stationary propagation medium in space, produce motional angles actually embedded in the medium, they are not apparent resolved angles. The motional angles are the same for the same speed but they are on each side of the normal to the direction of motion, distinguishing clearly between the two motions relative to the propagation medium. According to equations 8.6, the Earth's GFOD width W is about 50 times the Earth's radius. The GFOD is assumed to be spherical like, although there could be distortion through high forward speed motion. Further confirmation and detailed measurements of the medium velocity profile around the Earth and other planets in the solar system would be useful.

3.4 Electro-Gravitational Boundary Layer (EGBL)

Now consider the effect of body rotation within its GFOD. If the body has a massive atmosphere, refractive index $n>1$, the body's gravity appears to attract and drag the gravitational medium with its surface. This 'stationary' medium on the Earth's surface has been demonstrated by Sagnac's (1913) rotating mirrors producing propagation time asymmetry, Section 5.1 Chapter IV, and M&G (1925) fixed optical loop rotating with the Earth measuring increasing propagation speeds towards the equator, revealing the medium is clinging to and rotating with the planet's surface relative to the surrounding 'stationary' GFOD, described in Section 5.3, Chapter IV.

The medium acts as a compressible electrical fluid with an attractive coupling gradient $G_c=du/dy$, creating an electrical medium velocity gradient perpendicular to the body's surface. This appears to result in an Electro-gravitational Boundary Layer (EGBL) of effective width w, say defined by 90% of the maximum velocity in the y direction from the surface, (where effectively the rays cease to bend). For strong coupling gradient G_c is small, making w large.

Figure 8.4 uses radial ray propagation to illustrate the various possible boundary layer (BL) profiles. Figure (a) shows a rotating planet where the entire surrounding medium out to infinity is rotating with the planet, G_c=0, w=∞. Figure (b) depicts the planet motion relative to the surrounding medium completely at rest, zero coupling gradient G_c=∞, w=0, showing the source rotational angle:

$$(\theta_s)_{rot}=\tan^{-1}(M_{rot}), \quad M_{rot}=v_{rot}/c \qquad (8.10)$$

Figure (c) and (d) show intermediate steps between (a) and (b), i.e. with a finite BL of width w. Figure (c) shows a thin BL (steep velocity gradient) and Figure (d) a thick BL (less steep gradient).

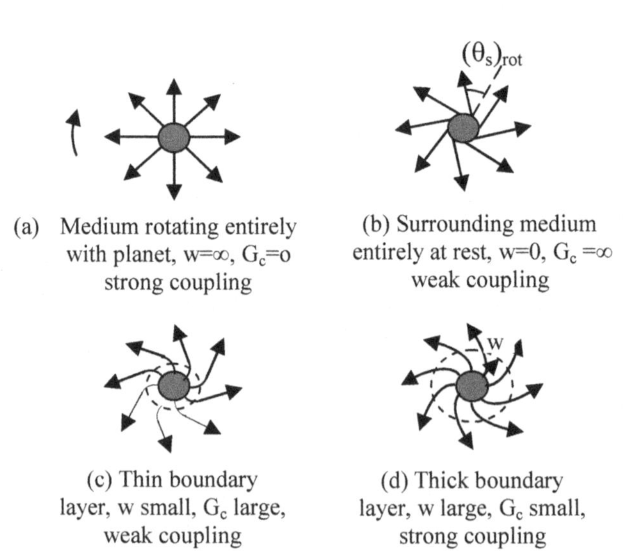

Figure 8.4 Rotating planets with respect to their 'stationary' GFOD medium, with various medium boundary layer profiles

Tangential rays at high frequencies, travelling initially in straight-lines along the planet's surface, would leave the planet. Whereas, low frequency radio waves, travel (bend) around in the propagation medium surrounding the planet's surface, as in early LF transmission between UK and Australia. According to H&K (1972) Section 5.7, Chapter IV, the EGBL extends to less than 10km above the Earth's surface. The Earth's surface velocity at the equator is 440m/s, giving an electrical attractive coupling gradient of $G \approx u_m/w \approx 440/10{,}000 \approx 0.044$m/s/m. More detailed measurements are required.

3.5 Global Positioning System (GPS)

GPS (1992) is calculated using triangulation of at least three positions from at least four local *artificial* satellites. For simplicity, an overhead geostationary satellite on the equator is assumed, rotating with the Earth, exaggerated in Figure 8.5. For a non rotating Earth, the propagation time between satellite (A) and Earth station (B) would be simply distance (h) divided by speed of light (c), i.e. $t=h/c$, there would be no side displacement. For an Earth and satellite rotating relative to the 'stationary' GFOD medium surrounding the Earth, the surface will rotate by $d_e=v_e t=v_e h/c=M_e h$ and the satellite by $d_s=v_s t=v_s h/c=M_s h$, in time t, where v_e and v_s are the velocity of the Earth's surface and satellite respectively.

The satellite signal in the medium will go from C-B (signal at B was emitted when the satellite was at A). For a zero boundary layer width w, propagating paths will be straight lines. For a finite boundary layer thickness, the propagation paths will be slightly curved. For small Mach numbers M_e, the Earth displacement d_e and the resulting satellite surface displacement angle $(\theta_s)_{sat}$ then become:

$$d_e \approx v_e h/c \approx M_e h \quad \text{and} \quad (\theta_s)_{sat} \approx \tan^{-1}(d_e/h),\ d_e/h \approx v_e/c \approx M_e \quad (8.11)$$

Lorentzian contraction (space and time), at moderate speeds, as well as the gravitational compression for moderate planet sizes, is small compared to the classical rotational displacement effect, Ashby (2003).

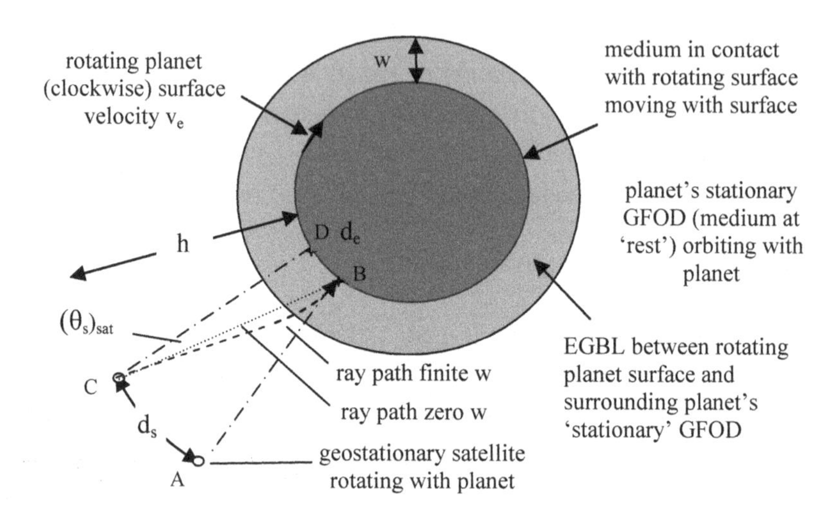

Figure 8.5 Global Positioning System (GPS). Satellite and Earth rotate relative to the surrounding 'stationary' GFOD medium causing motional angles and surface displacements that are predicted using classical theory. Relativistic effects are negligibly small at Earth speeds and small integration times (not to scale)

3.6. Displacement confirmation

At low speeds and vast distances from the planet, the effect of the local GFOD and the EGBL will be very small, but none the less close to the planet's surface both effects are important and measureable. For an Earth radius of $r_e \approx 6x10^6$m, rotation $\omega_e \approx 7.3x10^{-5}$radians/s, satellite radius $r_s \approx 26x10^6$m, height $h=r_s-r_e \approx 20x10^6$m, propagation time $t=h/c =20x10^6/3x10^8=66$ms and surface velocity $v_e=\omega_e r_e=7.3x10^{-5}x6x10^6=$ 440m/s, the Earth's Mach number at the equator becomes $M_e = v_e/c = (440/300)x10^6 \approx 1.5x10^{-6}$. The satellite angle and displacement are respectively, $(\theta_s)_{sat} \approx \tan^{-1}(M_e) \approx 10^{-4}$ degrees and $d_e \approx hM_e \approx 20x10^6x1.5x 10^{-6} \approx 30$m. This is the measured classical displacement. Over long periods of time corrections are needed for relativistic and gravitational time drifting, and for signal distortion through transmission through the ionosphere.

The satellite angle is obviously difficult to measure directly. But its displacement is measured routinely in GPS where it is important to measure position accurately. The displacement correction is automatically taken into account in the Global Positioning System (GPS) (1992), in calculating position. It is verified countless times daily, again confirming the presence of the propagation medium. The curvature of the propagation path, caused by the EGBL, is probably too small to measure. But the extent of the boundary layer of effective width w and its medium velocity profile are well within measurement accuracy, using the one way technique described in Section 6.1, Chapter I.

According to Einstein's Inertial Frame (EIF) (no medium, no propagation time difference between a stationary and moving frame), the propagation time for the satellite path A-B should be simply $t=h/c$. Here, the source ray angles and resulting displacement having no lateral motion should not exist ($(\theta_s)_{sat}=0$, $d_s=d_e=0$). i.e. Einstein etherless theory cannot account for these finite displacement angles and distances in transmission between satellites and ground stations. To account for the displacement, it should be acknowledged that the medium has to be taken into account and that the Earth and satellite rotate relative to the 'stationary' medium surrounding them.

3.7 Asymmetric signal delay

Apparently, EIF had given puzzling results previously to the GPS measurements. The effect of the propagation medium had caused scientists concern, through transpacific satellite relay links between USA and Japan, as measured by Saburi et al (1976) and analysed by Su (2001). Asymmetric propagation delays of the order of 300ns were unexpectedly found. These delays cannot be accounted for using Einstein's ether-less SR. The effect is exaggerated in Figure 8.6. Here the straight arrowed lines indicate the propagation paths, and e and r are the emission and reception position on the rotating Earth respectively. The slight curvature of the rays, through the boundary layer, is small at Earth speeds and is neglected. It can be seen that propagation, in the direction of the Earth's rotation, is longer than the propagation path against the rotation, relative to the surrounding stationary GFOD medium.

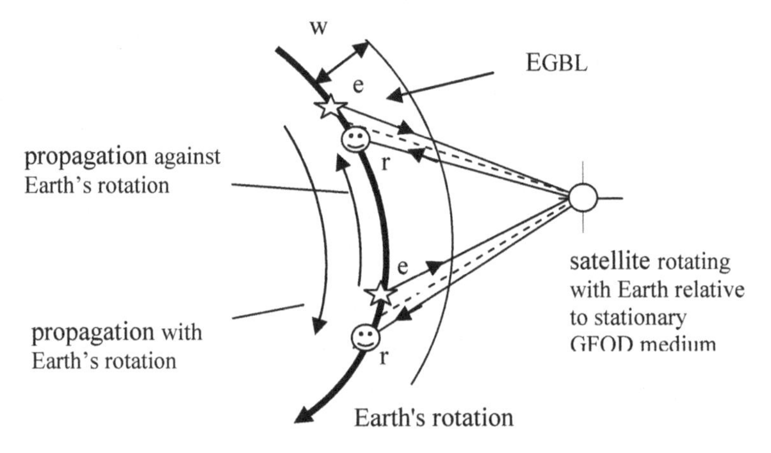

Figure 8.6 Satellite Propagation Time Asymmetry (PTA) caused through the satellite and Earth's rotation relative to surrounding stationary GFOD medium (not to scale)

If there was no propagation medium, as in EIF, or the medium moved completely with the Earth, as in Figure 8.4(a), then there would be no asymmetric propagation delay (the propagation paths would be equal in each direction as in the dotted line path). There is a measured asymmetric delay of 300ns ($300\text{x}10^{-9}\text{x}0.3\text{x}10^{9}\approx100\text{m}$), corresponding to the path difference, confirming the presence of the propagation medium. Again relativistic and gravitational effects, at these moderate speeds and Earth's gravity, are small in comparison to the classical displacement and propagation time delay. Satellite motional angles, signal displacements in GPS, and asymmetric propagation delays in transpacific satellite communications are a confirmation of the existence of the medium.

3.8 Planet's GFOD confirmation

The GPS and Saburi et al, verified calculations, depend on the medium surrounding the Earth being 'stationary' within the Earth's GFOD, orbiting with the Earth. If the GFOD did not exist, propagation would have been relative to the propagation medium at rest in 'space'. This would have caused a large Propagation Time Asymmetry (PTA) between the slowly rotating Earth surface speed and its considerable faster orbital speed (X100). These large PTA's have not been found in the M&G (1925), H&K (1972), Saburi et al (1976) or the GPS (1992) measurements. Thus these measurements confirms the presence of the Earth's GFOD, orbiting with the Earth.

Also details of the Sun's GFOD have been confirmed by Reasenberg et al (1979). Here well within the Sun's GFOD, interplanetary communications between solar planets have been predicted successfully considering the medium 'stationary' around the Sun. Measurements were made between Mars (Viking Lander) and the Earth. The first order effect of the rotation of the planets, their orbits around the Sun and the second order effect of the Sun's gravitational field on the wave propagation, were all taken into account. The success of the predicted communication path between Earth and Mars, relied on the fact that the propagation medium between the solar planets is basically 'stationary' within the Sun's GFOD, which surrounds the Sun and moves in the Milky Way at approximately 300 km/s ($M \approx 0.01$) through the universe. Again if the propagation medium had not moved with the Sun, then the Sun and Solar System's speed through the universe would have caused huge Propagation Time Asymmetries (PTA) in the forward and backward communication times between Mars and the Earth, which again have not been measured.

3.9 Medium and system motion

Based on the propagation medium, the following relationships between reference frames exist. In the Earth Centred Earth Fixed (ECEF) reference frame, as both the frame and medium rotate with the Earth, this frame gives predictions on the Earth's surface, in particular explaining the MMX (1887) and Sagnac's (1913) measurements. Away from the surface, the EGBL separating the rotating surface medium from the stationary surrounding medium is evident through M&G (1925). In the Earth Centred Inertial (ECI) frame, the frame moves with the Earth and medium in orbital motion around the Sun as in Hafele and Keating's (1972) measurements and satellite communications, Saburi (1976) and GPS (1992). In a Heliocentric frame the medium and frame move with the Sun and its Solar System, as confirmed in Reasenberg et al measurements. Outside the Sun's GFOD and the GFOD's of other suns and mass concentrations, inside and outside the vast spaces of the Milky Way and other galaxies, the propagation medium should be finally stationary, at rest on average, zero mean velocity, unless the medium is expanding appreciably.

It is difficult to study the presence of the medium across free space through measuring the source and observer motional event time transform, given by equation 6.2, i.e. $K_t=(\tau_o/\tau_s)=\varepsilon_s\ \alpha_s^{-1}\ \varepsilon_o^{-1}\alpha_o$. The problem here is at low speeds, only the difference velocity is possible as given by equation 5.29, Chapter V, not individual source and observer motions. The first part of equation 6.2 has been used to measure the relatively high velocities of distant stars, relative to the propagation medium, using the red shift, neglecting the Earth velocity. Here only the total motional and gravitational red shift can be measured, since it is difficult to separate the two without knowing the gravitation strength of the system under surveillance. A very heavy system could be mistaken as receding at high speed.

3.10 Measurements needed

The extent and gradient details of the Electro-gravitational Boundary Layer (EGBL) on or above a planet's surface, and the extent and gradient details of the medium surrounding and moving with the planet's Gravitational Field of Dominance (GFOD), both identified in Section 6, Figure 2.8, Chapter II, and confirmed in this Section, should be investigated in detail. The local medium velocity profiles and boundary layer width w, could be mapped using tethered balloons or low and high altitude aircraft. This could be achieved by measuring classical Propagation Time Asymmetry (PTA) using laser interferometers, or relativistic time slowing using atomic clocks.

The medium's velocity at the equator relative to the stationary medium surrounding the Earth is s=460m/s ($M=s/c=460/3x10^8 \approx 1.5x10^{-6}$). According to H&K, Section 5.7 Chapter IV, the medium's velocity is close to zero at 10 km above the Earth's surface relative to its GFOD. This profile needs to be mapped in detail. Using one way PTA measurements for b=3m, $\Delta t \approx Mb/c=1.5x10^{-6}x3/3x10^8 \approx 1.5x10^{-14}$ s, or $N=c\Delta t/\lambda$ = 7.5 interference bands. Or using relativistic time slowing and atomic clocks over a 24 hour period ($8.6x10^4$s) gives $\Delta\alpha \approx -M^2/2 \approx -1.5^2x10^{-12}/2 \approx -1.1x10^{-12}$, and $\Delta\tau_s \approx \Delta\alpha\tau_p \approx -1.1x10^{-12}x8.6x10^4 = -95$ns /day. Atomic clocks are sensitive to gravity, but PTA measurements are not.

The medium gradient details and extent of the Earth's effective GFOD radius, (R_{GFOD}), could be investigated using space vehicles or satellites at various distances from the Earth. These could be compared with measurements across the Moon's and Sun's GFOD. Perhaps laser beams could be tracked as they pass and bend through the EGBL and the GFOD gradients from various directions. Measurements around other planets, of various sizes, in the Solar System could also be investigated to establish, for example, the effective EGBL extent w above the planet's surface compared to its effective GFOD extent (R_{GFOD}), around the planet. A measurement of w/R_{GFOD}, as a function of the planet's gravitational mass in the presence of other gravitational fields would be helpful.

4 Conclusions

The Atomic Residual Difference Field (ARDF) between distributed dissimilar charges within atoms (mainly electrons and protons) appears to be the origin of gravity and the attraction between matter. The total gravitational matter of the universe, through its residual gravitational field, appears to provide the Universal Gravitational Reference Field (UGRF) at rest in space. This field appears to create mass inertia and provides an absolute reference for motion.

Considerable data confirms that away from gravitational bodies the propagation medium on average is at rest. Close to moving heavy bodies (planets), the propagation medium appears to be attracted to and moves with the body, within the body's Gravitational Field of Dominance (GFOD). On the surface of a rotating body, the propagation medium can rotate with the surface, producing an Electro-gravitational Boundary Layer (EGBL) between the rotating surface and its 'Stationary' GFOD medium. The EGBL has been confirmed through MMX, Sagnac, and M&G.

The GFOD has been established through Propagation Time Asymmetry (PTA) delays in satellite communication systems (Saburi et al), signal displacement on Earth, in Global Positioning Systems (GPS), lack of PTA in interplanetary communications (Reasenberg et al), and through the symmetrical time slowing moving relative to the Earth's GFOD medium, (H&K). Further details of the EGBL and GFOD should be investigated, their extent, gradients and their asymmetry if any, with motion, for different planet sizes within the Solar System, compared to that of the Sun and elsewhere.

Chapter IX:

Concluding Chapter

1 Introduction

The investigation has been concerned with the following aspects of Einstein's Special and General Relativity.

1. It has been established that the propagation medium (ether) for the transmission of light exists. That there is sufficient evidence showing the vacuum medium, away from gravitational mass, is generally at rest in free space, providing an absolute reference.

2. The medium has finite electrical properties. Permeability (inertia) and permittivity (stiffness), but no mass or rigid (atomic) structure. It therefore acts as a mass-less deformable electrical fluid medium.

3. The medium is attracted and compressed by gravity and contracts structures passing through the medium, both time and space.

4. It has unity refractive index (cannot convect light). But can refract (bend) light through medium compressions and velocity gradients.

5. A Gravitational Entrainment Model (GEM) is described where the medium orbits with gravitational bodies within their Gravitation field of Dominance (GFOD). The medium can rotate with the body forming an Electro-gravitational Boundary Layer (EGBL) on its surface relative to the surrounding stationary medium.

6. The medium transmits an electric Atomic Residual Difference Field (ARDF), from a finite distributed of dissimilar charges within atoms and molecules, attracting similar distributions, creating gravity.

7. A residual ARDF from all gravitational matter in the universe provides a Universal Gravitational Reference Field (UGRF) with no net force, but potential energy creating inertia for mass in motion.
8. An EM Motional analysis (EMMA), for the first time, describes a complete process of radiation, propagation and reception between EM sources and observers in motion.
9. Einstein's ether-less concept of relativity is a simulated effect, represented by oblique transform axes, rather than Lorentz's rectangular medium based ones. All motional effects, both classical and relativistic, are relative to the propagation medium, not relative between systems, as Einstein believed.
10. Einstein's measured properties in his SR and GR, although claimed to be ether-less, are in fact predicted using field equations based on Maxwell's medium and Lorentz's medium based transform.

2 Data Supporting Medium

Each reference below, which is justified in detail in the main text, is supportive of the propagation medium. The data basically shows that the propagation medium is generally at rest in space or moves with gravitational bodies. In the following list, in chronological order, C indicates classical (Galilean, non-relativistic) theories, which contain the classical medium based Propagation Time Asymmetry (PTA). R includes the additional relativistic (Lorentzian,) system time and structural contraction, passing through the medium.

1. Bradley (1725), showed that pointing a telescope slightly in front of a star to enable it to be viewed (stellar aberration angle), is caused by the observer (telescope) on Earth moving relative to the star light propagating in the medium at rest in 'space' (C).

2. Fresnel (1818), and later experimentally Fizeau (1851), showed that a moving dense medium, refractive index n>1, can convect light propagating in a vacuum medium, increasing its propagation speed in the direction of motion (R).

3. Maxwell (1865) derived and solved the EM wave equation predicting causal EM disturbances (light) propagating relative to a propagation medium of permittivity (electrical inertia) μ and permeability (electrical stiffness) ę (C).

4. Michelson and Morley (1887) showed that there was no Propagation Time Asymmetry (PTA) in and perpendicular to the direction of the Earth's motion passing through the universe, indicating that the medium is at rest relative to (moves with) the Earth (C).

5. Lorentz (1899), predicted that time and space contract by exactly the same ratio, through motion with respect to the medium, maintaining the propagation speed as invariant in the moving frame. Also that the Propagation Time surrounding a moving system is Asymmetric and variant (R).

6. Sagnac (1913), through his rotating mirrors on Earth, established PTA relative to the stationary propagation medium on Earth. It was predicted exactly according to the classical part of Lorentz's wave theory (C).

7. Michelson and Gale (M&G) (1925), using a large optical loop, stationary on the Earth’s surface, showed that the light speed increased towards the equator according to the Earth's surface speed, showing that the medium rotates with the Earth’s surface, relative to a ‘stationary’ medium surrounding the Earth (C).

8. Items 6 and 7 infer that a 'stationary' medium orbits with the Earth within the Earth's Gravitation Field of Dominance (GFOD), in the presence of the Sun's GFOD. An Electro-gravitational Boundary Layer (EGBL) between the rotating Earth's surface and its 'stationary' GFOD is formed (C).

9. Cerenkov (1934), demonstrated that a high speed electron, leaving a vacuum and entering a more dense, lower propagation speed medium (air), produces a medium based 'super lightic' boom (shock wave), similar to a medium based supersonic boom in air, or a ship's surface bow wave in water (C).

10. Penzias and Wilson (1965), detected the EM cosmic microwave background (CMB) radiation left over from the big bang, which propagates omni-directionally throughout the universe relative to the propagation medium at rest in space (C).

11. Hafele and Keating (H&K) (1972), using atomic clocks, showed that an aircraft ages symmetrically according to its speed with respect to the 'stationary' medium surrounding the Earth. (Not relative to the rotating Earth's surface, where time is asymmetrical.) This confirms the existence of the Earth's stationary propagation medium within its GFOD (R).

12. Saburi et al (1976), through a Satellite Communications link across the Pacific between the USA and Japan measured a 300ns PTA, transmitting in and against the direction of the Earth's rotation with respect to the surrounding 'stationary' medium, again confirming the Earth's 'stationary' medium within its GFOD (C).

13. Picasso et al (1977), through measuring the half life of an orbiting high speed muon, showed that its ageing slows

moving relative to the medium on the Earth's surface, again confirming the medium's 'stationary' presence on Earth (R).

14. Brecher (1977), argued that orbiting EM sources (binary stars) move relative to the propagation medium at rest in space. Otherwise the medium moving with the star in the far field would produce gaps and overlapping radiation on retreating and advancing flight paths, which is not measured (C).

15. Reasenberg et al (1979), in the Earth-Mars/Viking Lander project, confirmed the presence of the Sun's GFOD, through no PTA in signal communication propagation forward and backwards relative to a the 'stationary' medium moving with the Sun and solar system (C).

16. GPS (1992), demonstrated backward displacement source angles and PTA, resulting in positional displacement of 30m on the Earth's surface, caused by the satellite and Earth rotating relative to the Earth's GFOD. Again confirming the 'stationary' medium surrounding the Earth. If the GFOD had rotated with Earth there would have been no displacement (C).

17. COBE (1992), moving in any direction through space relative to the cosmic microwave background (CMB), causes increased radiation collection confirming the propagation medium transmitting the CMB is at rest in space. Similar to trawling nets, in any direction, catch more fish than stationary ones (C).

3 Data Interpretation

It can be seen that there is considerable evidence in support of the vacuum medium. 13 out of the above 17 references are classical (Galilean) effects. The first 9 were available in Einstein's time. The remaining 4 out of the 17 are supported by the relativistic (Lorentzian) influence, which further reaffirms the existence of the medium. COBE confirmd that CMB radiation in the universe propagates with respect to this stationary medium. Bradley and Brecher demonstrated that EM observers and sources move with respect to this stationary medium.

Sagnac's PTA showed that his mirrors rotated relative to the 'stationary' medium on the Earth's surface. Michelson and Gale's (M&G) measured increased light speed towards the equator, demonstrated that the medium on the Earth rotates with the planet's surface, relative to a 'stationary' medium surrounding the Earth, implying a medium boundary layer between the rotating surface and the surrounding stationary medium, within the Earth's GFOD.

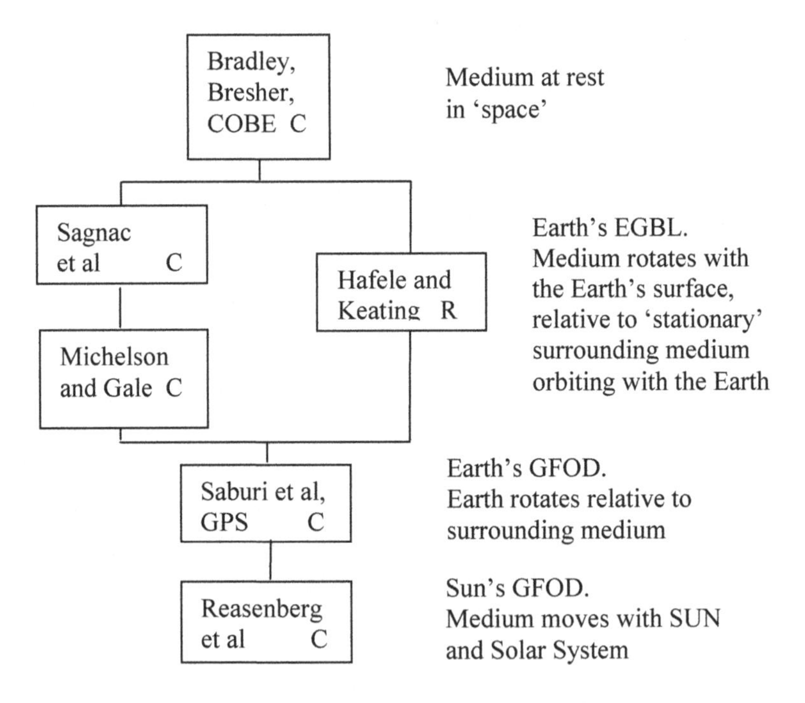

Figure 9.1 Justification for Gravitational Entrainment Model (GEM) shown in Figure 2.3. C-classical, Galilean propagation time change (PTA) through motion. R-relativistic, Lorentzian space and time contraction through motion

Hafele and Keating (H&K) showed that a high flying aircraft's ageing depends on its velocity relative to the 'stationary' GFOD medium surrounding the Earth, not relative to the rotating Earth's surface. These six references demonstrate that the medium is generally at rest in space, orbits with the planet, rotates with the planet's surface, creating an Electro-gravitational Boundary Layer (EGBL) immediately above its surface. This supports a Gravitational Entrainment Model (GEM) considered initially in Section 2 and 3, Chapter I and in detail in Section 6, Figure 2.8, Chapter II, and summarized in Figure 9.1.

Thus i) Bradley, Brecher and COBE support propagation with respect to a medium at rest in 'space'. ii) Sagnac supports propagation with respect to a medium rotating with the Earth's surface and iii) M&G and H&K show that the surface medium rotates relative to a 'stationary' medium surrounding and orbiting with the Earth. Therefore, there is ample confirmation for both a propagation medium at rest in 'space' and one that moves with gravitational bodies. Further, it is not useful to deny the existence of the propagation medium, particularly when the medium has well defined and measurable properties, described in Section 2, Chapter IV.

It appears that the medium orbits with the Earth within its Gravitational Field of Dominance (GFOD) in the presence of the Sun's gravitational influence. This is supported through the measurements of M&G, H&K, Saburi et al and the GPS. The details of the first two are considered in Section 5, Chapter IV. And the last two satellite references are considered in Section 3, Chapter VIII. If the surrounding medium had not orbited with the Earth, then the propagation would have been with respect to the medium at rest in 'space'. This would have caused large PTA in the satellite signal transmissions, in and against the direction of the Earth's motion through 'space', which has not been measured. If the GFOD had rotated with the Earth there would have been no PTA in the Saburi et al satellite communications and no displacement in the GPS satellite motions.

Further, Reasenberg et al, have confirmed through no PTA that the medium moves with the Sun and its GFOD, through solar inter-planetary communications. It is not possible to explain these moving media phenomena without a medium. The medium's presence, which does not support Einstein's ether-less relativity, could have been established a long time ago using just classical measurements, avoiding the setback created through Einstein's ether-less interpretation of relativity in 1905.

4 Revised Optical Process

No medium is not an option in wave theory. Its lack cannot define a causal description of time and space. There is no reference in which to judge motion or the direction of time. Its wave equation cannot be solved, causal predictions cannot be made. Whereas, restoring the medium, time and space become well defined. It then becomes clear exactly what happens and when. Defining time to pass at a constant rate, away from gravity and motion, then enables a sequence of events and rate of happening to be observed and judged relative to that rate, for systems in motion or in gravitational fields. Source motion through the vacuum medium, according to Lorentz, contracts both time and space, gravity according to Schwarzschild, compresses the medium's time and space, and observer motion expands time and space, making the connection between gravity and acceleration.

Universal Gravitational Reference Field (UGRF), from the ARDF from gravitational matter in the universe, from all directions in space, creates a resultant zero intensity field. However it has a finite scalar potential, appearing to create an inertial field apposing mass in acceleration. This field explains why rotating planets bulge at their equator, Foucault's pendulum oscillates independently of the Earth's rotation, spinning tops stay upright and travellers develop motion sickness. They all appear to depend on motion with respect to the residual field. Without the medium, this field and none of these effects above could occur. Assuming the universe is not expanding too rapidly, the medium in space, away from gravitational matter, can be considered to be at rest throughout the universe. It provides an absolute reference for systems passing through the medium.

Rotating masses could have a similar mechanism to matter spiralling around and being pulled into intense rotating gravitational fields (black holes). However, close to a moderately dense rotating planet (small event horizon), the effect is small, resulting in only the medium being attracted and orbiting with the planet. Rotating planets appear to rotate the propagation medium close to their surface forming a medium boundary layer with the 'stationary' medium surrounding the planet.

Lens/Thirring (1918) suggested that rotating gravitational matter could create a couple (frame drag) on nearby bodies. The cause is thought to be a Gravito-magnetism (GM) effect, where rotating bodies are considered to generate a gravitational effect analogous to a moving charge generating a magnetic field. This effect is considered very small indeed, approximately one trillionth compared to regular gravity. It is indeed a challenge to measure the GM effect on satellite orbits and gyroscopes, as attempted by the Gravity Probe B (2011). However, medium rotation and velocity profiles close to planet surfaces, should be much easier to measure.

Figure 9.2 shows diagrammatically, the Earth orbiting around the Sun, illustrating its Gravitational Field of Dominance (GFOD) of width W and its Electro-gravitational Boundary Layer (EGBL) of width w. From equation 8.6, a simple estimate of the ratio $W/D \approx 1/500$ and $r/W \approx 1/50$ for the Earth is calculated, where D is now the distance of the Earth from the Sun, and r is the radius of the Earth. The EGBL width w is indicated to be a small fraction of the Earth's radius. According to Hafele and Keating (H&K) (1972) measurements, w extends to less than 10km above the surface. Also shown in Figure 9.2

is time slowing through motion and gravity calculated from equations in Section 7 Chapter I and in Section 3.2, Chapter VIII or in detail in Section 7, Chapter V.

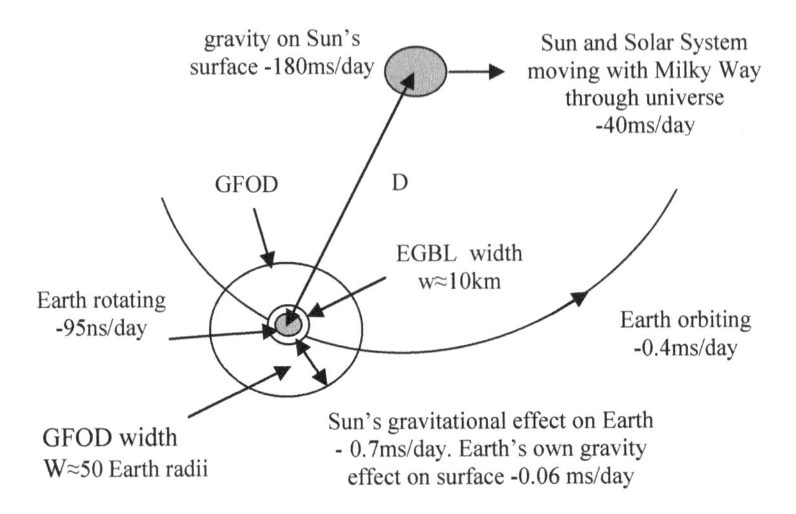

Figure 9.2 Earth orbiting around the Sun illustrating its Gravitational Field of Dominance GFOD of radius W and the Electro-gravitational Boundary Layer EGBL of width w. Also shown is time slowing through motion and gravity

The time contractions through motion are: Sun and Solar System rotating with the Milky Way through the universe with velocity 300km/s ($M \approx 10^{-3}$) giving -40ms/day. Earth orbiting around the Sun ($M \approx 10^{-4}$) giving -0.4ms/day (≈10x less slower in motion than the Milky Way, but 100x less time slowing). Earth rotating at the equator ($M \approx 1.5.10^{-6}$) slows by -95ns/day. The gravitational compressions slowing time on the Sun's surface are -180ms/day, and on the Earth through the Sun's presence is -0.7ms/day

It is inconceivable infield theory, or any other logical physics, for EM waves to be received before they are emitted (non causality). Or propagation to occur without a propagation medium. Such a universe, without a medium, would require some other form of metaphysical process to transmit light, otherwise there would be complete darkness. It appears that Einstein's ether-less relativity, based on no medium, is an unfortunate description, created through simulation, using oblique transform axes, rather than reality represented by the rectangular medium based axes used by Lorentz. Einstein's ether-less aspect of relativity is non causal, it cannot be measured, it is not reality.

The notion of relative motion between two objects affecting time and space directly, which Einstein proposed, has been accepted for over a century. It is a concept which has no physical foundation. It is believed therefore, that it is very important to establish a clear distinction between hard verifiable physics (reality), and unproven modelling. Otherwise, without experimental confirmation of models, it is difficult to distinguish between science and science fiction.

5 Gravitational Attraction

The only static attractive force known in nature, apart from nuclear, is the electrical force between charges. Van der Waals (1873) and London (1937) established that molecules attract each other in the near field. This is through permanent electric dipoles (dissimilar charges), and induced dipoles through electron orbit displacement. Gravity, considered in Section 2, Chapter VIII, can be considered as a far field extension to the near field atomic attraction. The difference electric field between dissimilar charges from a large assembly of atoms and molecules constituting gravity. Although a point distribution of atoms and molecules with equal numbers of positive and negative charges is neutral, a finite distribution (non point) produces a field that does not quite cancel, leaving a weak but finite attractive field.

This difference electric field has similar properties to gravity, it is attractive, it requires a medium with a retarded propagation time and inverse square law decay with distance. Attractive mass therefore can be considered as a difference electrical field from a finite distribution of dissimilar charges within atoms and molecules. It has a weak charge equivalent of 10^{-20}Coulombs. All electric fields appear to use the same electric medium to transmit their steady state values and disturbances. Unsteady electrical fields create EM waves, steady difference electric fields create gravity, and the residual gravity field from all the gravitational mass in the universe creates the inertial field.

6 Einstein's Ether-less Belief

Einstein believed that there was no propagation medium required to transmit EM waves and gravitational fields. Whereas, it is found that:

1 All wave theories require a medium to propagate their waves, solve their wave equation and give causal (predicable) results. Although Einstein (1905) believed that there was no medium (ether), he used one in his field equations to predict his causal observations.

2 There is nothing special regarding electromagnetic (EM) waves and gravity, they both require a medium, as do all field theories. The only difference between classic and EM wave theories is that time and structure shrink passing through the medium, and gravity compresses the medium's time and space, neither are possible without a propagation medium.

3 EM motional theory can be derived directly from a medium based classical theory by simply modifying according to the medium based Lorentz (1899) time and space contraction through motion. Lorentz confirmed the preferred frame of reference, which Einstein attempted to remove simply to explain the results of the invariant Michelson and Morley

Experiment (MMX)(1887). The MMX is now explained quite naturally with the propagation medium moving with the Earth.

4 Medium based concepts, used to support oblique transform axes, simulating non preferred ether-less motional properties, cannot be used to discredit the existence of the propagation medium (circular argument). Whereas, Lorentz's rectangular axes medium based transform predicts the causal observations, and confirms the preferred reference frame.

5 It is not possible to use relativistic arguments to explain basic EM motional effects at Earth speeds, eg. Sagnac (1913) and GPS (1992). In these examples relativistic effects are small requiring a large summation period. Whereas, the classical Propagation Time Asymmetry (PTA), based on the propagation medium, is the dominant instantaneous motional cause.

6 Einstein's ether-less relativity, based on his inertial frame, is non causal, it cannot predict measured observations. To observe events from Einstein's inertial frame requires the propagation medium to be restored to allow the events to reach the observer.

7 Einstein's claim that his inertial frame is absolutely invariant, is false. Only the mechanics and speed of light are invariant in the moving frame. The effect of the propagation medium creates a PTA upstream and down, in the moving frame, making the predictions causal and variant, the same as all causal wave theories.

8 Allowed solutions of the wave equation are that the medium can move with the frame where there is no PTA, or there is motion with respect to the medium where there is PTA. But the propagation defined through the causal solution of the

wave equation must always be relative to the medium, not relative to the moving frame as Einstein believed.

9 This is Einstein's basic error, the propagation is always relative to the medium. The propagation speed remains invariant in the moving frame only because both time and space contract by exactly the same ratio through motion with respect to the medium, maintaining its speed.

10 Einstein believed that no PTA in the Michelson and Morley Experiment (MMX) (1887) supported his inertial frame, whereas in fact it is shown to support medium motion with the Earth. Einstein's SR (1905) and GR (1915), which were thought to be ether-less are shown in fact to be medium based.

7 Conclusions

A new relativity (NR) theory removes ambiguities and paradoxes present in Einstein's relativity. NR is based on Maxwell's (1865) medium, it is an extension of Lorentz's (1899) motional theory for both source and observer motion, which Einstein's ether-less aspect of Special Relativity (SR) (1905) cannot include. NR is therefore an extended version of Einstein's medium based aspect of SR. It predicts the same measured predictions as SR. It also predicts other measured observations (source and observer motional differences) that SR cannot predict, and shows that all other ether-less predictions are false (non causal). In brief it is found that:

1. All waves require a propagation medium to propagate through space and to be predictable (causal).
2. EM waves are no exception, they require a propagation medium, according to Maxwell.
3. The same EM medium propagates light, gravity and the inertial field.

4. Einstein's measured SR predictions are medium based (causal). his ether-less predictions are non causal (cannot be measured).
5. SR is incomplete it does not distinguish between source and observer motion, or between stationary and moving media.
6. A new medium based theory 'EMMA' is developed from the Lorentz transform accounting for item 5.
7. The new theory unites classical and modern physics, and electro-magnetism and gravity.

Einstein's medium based aspect of SR, which predicts many of today's measured observations, is based on Lorentz's medium based motional theory. Einstein's ether-less aspects of SR (simultaneity, optical reciprocity, time travel and no absolute time and space),cannot be measured. They are the result of a mathematical simulation equivalent to an oblique axes transform. This oblique axis transform is not a solution of the wave equation, it is non causal and not supported by the medium based rectangular axes Lorentz transform.

No data or causal arguments have been made that justify this oblique transform. To restore reality and propagation with respect to the medium, the oblique axes are returned back to their regular rectangular Lorentzian positions. This breaks the propagation time symmetry, removes the non measurable simulated concepts of simultaneity and reciprocity, predicted by Einstein's invariant inertial frame, and restores the variant Propagation Time Asymmetry (PTA) laid down in the stationary propagation medium by moving systems.

There is no basis for the ether-less aspect of SR, apart from attempting to satisfy the Michelson and Morley Experiment null result (MMX) (1887), which was judged to support an ether-less universe. It is now accounted for quite naturally through the medium on the Earth's surface rotating relative to the stationary medium surrounding and orbiting with the Earth. This creates an Electro-gravitational Boundary layer (EGBL) between the rotating surface and the surrounding stationary medium. Michelson and Gale's (1925) fixed optical loop showed that the surface medium rotates relative to the stationary medium surrounding the Earth. Also Hafele and Keating's (1972) flying atomic clocks 10 km above the Earth confirmed the medium at this altitude was stationary relative to the Earth's axis (not rotating).

The EGBL moving with the Earth not only accounts for the MMX, but also and Bradley's (1725) stellar aberration angle. It was thought that the aberration angle was a resolved apparent angle between the Earth's orbital motion and the speed of light and that the medium surrounding the Earth would affect it. However, the vacuum medium having unity refractive index, has no ability to convect light. The angle is actually an embedded angle in the medium, unaffected by the medium's presence. It provides a smooth transition between the medium at rest in space and the medium surrounding and orbiting with the Earth. Being a compressible electrical fluid, the medium can refract light around gravitational bodies, through gravitational compressions and medium velocity gradients, caused by the body's motion though the propagation medium.

Although Einstein denied the medium's existence and creatively described elaborate details of an ether-less SR, such as time travel and no absolute time and space, these details have not been measured. Einstein actually used a medium in his motional electro-dynamics field equations, referring motion to the medium rather than relative motion between systems. This violated his own concept of relative motion, rendering his ether-less SR model as untenable. Einstein's space in his space-time used to measure distance in his SR, and space time distortion (medium compression both time and space through gravity, according to Schwarzschild (1916)), are in fact the same space used by Lorentz (1899), filled with his medium. Measured system time and structure contraction, caused through motion relative to the propagation medium, supports the medium. The same medium is used in accelerating observer frames, linking these concepts together.

NR is in agreement with and supported by the transverse Doppler, changes in time, space, mass, momentum and Einstein's famous energy equation. Also, it is in agreement with and supported by Schrödinger's (1927) standing wave equation, Relativistic Quantum Mechanics (RQM), which is Lorentz invariant, the theories of Dirac (1928), Quantum Electro Dynamics (QED) and the Standard Model (SM) in particle physics.

Einstein's GR which results through expressing the LT in terms of Minkowski's (1908) space-time four-vector analysis and his gravitational theory, which form the basis of astrophysics and cosmology, are all based on the same medium. It is shown that the medium supports the Schwarzschild metric and event horizon. Assuming the medium to be homogeneous and isotropic, the Robertson(1936) and Walker (1937) metric in cosmology is obtained, and through the Friedmann metric (1924), the Hubble constant (1929).

It is believed that it has been shown beyond doubt that the EM medium (ether) exists, and that it is false to believe that electric fields and their disturbances can propagate through space without a propagation medium. Without the medium none of the motional effects discussed in this book could occur. There is no need to remove the propagation medium as Einstein attempted to do, the universe works extremely well with it. This new theory is not remarkable; it simply restores the rationality of EM theory by re-establishing its medium.

This in turn restores the connection between classical and modern physics. The confirmation that the medium exists, and that electromagnetic, gravitational and inertial fields all appear to be forms of an electric field, using the same medium, now provides a common basis for a unification theory of the universe. Finally 'ether' is preferred, as 'aether' usually implies its demise. Further details are available at www.new-relativity.com or author contact through ecasssw@outlook.com.

References

Galileo 1632. Galilei .*Dialogue Concerning the Two Chief World Systems* using the example of a ship travelling at constant velocity.
Huygens 1678. Christiaan Huygens, *Traité de la lumiere* Leiden, Netherlands: Pieter van der Aa.
Newton 1687. Isaac Newton Philosophiæ Naturalis Principia Mathematica. (published in Latin in 1687; revised in 1713 and 1726; and translated into English in 1729).
Bradley 1725. James Bradley. Third Astronomer Royal. Reprint from the Philosophical Transactions of Stellar Aberration: On the Motion of the Fixed Stars.
Young 1804. Thomas Young, Experimental Demonstration of the General Law of the Interference of Light, "Philosophical Transactions of the Royal Society of London", vol 94
Fresnel 1818. Lettre d'Augustin Fresnel a Francios Arago sur l'influence du movement terrestre dans quelque phenomenes d'optique. Annales de chimic el de physique 9 57-66.
Doppler 1842. Christian Doppler. On the coloured light of the binary refracted stars and other celestial bodies - Attempt of a more general theory including Bradley's theorem as an integral. Monograph.
Faraday 1844. Michael Faraday (1844). *Experimental Researches in Electricity*. **2**. Also Michael Faraday (1859). *Experimental Researches in Chemistry and Physics*.
Stokes 1845. George Gabriel Stokes. On the aberration of light. Philosophical Magazine 27: 9-15.
Foucault 1850. Foucault, M. L. Physical demonstration of the rotation of the Earth by means of the pendulum, Franklin Institute, 2000, retrieved 2007-10-31. Translation.
Fizeau 1851. Fizeau M H. "Sur les hypothèses relatives à l'éther lumineux". *Comptes Rendus* **33**: 349–355. Also 1859. *Ann. De Chim. Et de Phys.* **57**: 385–404.
Maxwell 1865. James Clerk Maxwell. A dynamical theory of the electromagnetic field. Philosophical Transactions of the Royal Society of London 155, 459-512.
Van der Waals, 1873. JD van der Waals *Over de Continuiteit van den Gas- en Vloeistoftoestand (on the continuity of the gas and liquid state)*. PhD thesis (excerpt), Leiden, The Netherlands.

Michelson and Morley 1887. Michelson, A. A.; Morley, E. W. On the relative motion of the Earth and the luminiferous ether. American Journal of Science 34 (203): 333–345.
Mach 1887. Ernest Mach, Mach's Principle Mach's archive Deutches Museum Munich.
Hertz 1888. Heinrich Hertz. Ueber sehr schnelle electrische Schwingungen, *Annalen der Physik*, vol. 267, no. 7, p. 421-448.
Fitzgerald 1889. The scientific writings of the late George Francis Fitzgerald. by J Larmor, Dublin, 1902.
Larmor 1897 Larmor J. On a Dynamical Theory of the Electric and Luminiferous Medium, Part 3, Relations with material media *Phil. Trans. Roy. Soc.* **190**: 205–300.
Thompson 1897. Joseph John Thomson, "Cathode Rays". *The Lon don, Edinburgh, and Dublin Philosophical Magazine and Journal of Science* (Bristol: Institute of Physics Pub.) Fifth Series: 296.
Lorentz 1899. Hendrik Antoon Lorentz. Simplified theory of electrical and optical phenomena in moving systems. Proc. Acad. Science Amsterdam 1: 427-442.
Poincaré 1900. Henri Poincaré. La théorie de Lorentz et le principe de réaction, *Archives Néerlandaises des Sciences exactes et naturelles* **v**: 253–278.
Planck 1900. Max Planck. "On the Law of Distribution of Energy in the Normal Spectrum". Annalen der Physik, vol. 4, p. 553.
Einstein 1905. Albert Einstein. On the electrodynamics of moving bodies. Annalen der Physik, 17: 891-921.
Minkowski 1908. Herman Minkowski "Raum und Zeit
Ritz 1908. Walter Ritz Translated from Recherchs critiques sur l'Électrodynamique Générale, *Annales de Chimie et de Physique,* Vol.13, p 145.
Bohr 1913. Niels Bohr. "On the Constitution of Atoms and Molecules, Part I". *Philosophical Magazine* **26**: 1–24.
De Sitter 1913. Willem de Sitter. "An astronomical proof for the constancy of the speed of light" Trans from *Physik. Zeitschr.* **14**, 429.
Sagnac 1913. Georges Sagnac. L'éther lumineux démontré par l'effet du vent relatif d'éther dans un inter-féromètre en rotation uniforme. Comptes Rendus 157, S. 708-710. Georges Sagnac. Sur la preuve de la réalité de l'éther lumineux par l'expérience de l'interférographe tournant. Comptes Rendus *157, 1410-1413.*

Einstein 1915. Albert Einstein. Die feldgleichungen der gravitation (The field equations of gravitation), Koniglich Preussische Akademie der Wissenschaften: 844–847.
Schwarzschild 1916. Über das Gravitationsfeld eines Massenpunktes nach der Einstein'schen Theorie. Sitzungsberichte der Königlich Preussischen Akademie der Wissenschaften **1**, 189-196.
Lense/Thirring 1918. Lense, J.; Thirring, H. (1918). "Über den Einfluss der Eigenrotation der Zentralkörper auf die Bewegung der Planeten und Monde nach der Einsteinschen Gravitationstheorie". *Physikalische Zeitschrift* **19**: 156–163. On the Influence of the Proper Rotation of Central Bodies on the Motions of Planets and Moons.
Kaluza 1921. Theodor Kaluza, "Zum Unitätsproblem in der Physik". *Sitzungsber. Preuss. Akad. Wiss. Berlin.* 966–972.
Friedmann 1924. Friedmann, Alexander Über die Möglichkeit einer Welt mit konstanter negativer Krümmung des Raumes, *Zeitschrift für Physik A* **21**: 326–332.
DeBroglie 1924. L. de Broglie, *Recherches sur la théorie des quanta Ann. Phys.* (Paris) **3**, 22.
Born 1924. Max Born, Einstein's theory of relativity. First published in 1924 available through Dover, New York, 1964.
Miller 1925. Dayton Miller, "Ether-Drift Experiments at Mount Wilson", *Proceedings, Nat. Acad. Sciences*, 11:306-314.
Michelson and Gale 1925. Albert Abraham Michelson, Henry G. Gale. The effect of the Earth's rotation on the velocity of Light. The Astrophysical Journal 61 S. 140-145.
Heisenberg 1927. W. Heisenberg, *Über quantentheoretishe Umdeutung kinematisher und mechanischer Beziehungen, Zeitschrift für Physik*, **33**, 879-893,
Schrödinger 1927. Erwin Rudolf Josef Alexander Schrödinger Annalen der Physik "*Quantisierung als Eigenwertproblem*".
Dirac 1928. Paul Dirac The Quantum Theory of the Electron. *Proceedings of the Royal Society of London. Series A,* **117** (778): 610–624.
Michelson, Pease and Pearson 1929. Michelson, A.A., Pease, F.G. & Pearson, F.: "Repetition of the Michelson-Morley Experiment", *Nature*, 123:88, 19 Jan. 1929; also in *J. Optical Society of America*, 18:181,
Hubble 1929. Edwin Hubble. "A relation between distance and radial velocity among extra-galactic nebulae". *PNAS* **15** (3): 168–173.

Kennedy and Thorndike 1932. R. J. Kennedy and E.M. Thorndike, Experimental Establishment of the Relativity of Time, *Phys. Rev.* 42 400-418.
Cerenkov 1934. Cerenkov, P.A., "Visible emission of clean liquids by action of γ radiation", *Doklady Akad. Nauk SSSR* 2 451.
Robertson 1936. Robertson Howard Percy. Kinematics and world structure, *Astrophysical Journal* 83 p 257.
Walker 1937. Walker Arthur Geoffrey. "On Milne's theory of world-structure", *Proceedings of the London Mathematical Society 2* **42**: 90–127.
London 1937. "The general theory of molecular forces", *Transactions of the Faraday Society* **33**: 8–26.
Turner and Hill 1964. K.C. Turner and H.A. Hill (1964). New experimental limit on velocity-dependent interactions of clocks and distant matter. Phys. Rev. 134, 252-256. 13 April 1964.
Higgs 1964. Higgs P.W. Brocken symmetries, massless particles and gauge fields. Phys. Lett. 12, 132.
Fox (1965). J G Fox. Evidence against emission theories. American Journal Physics Vol 3(1).
Penzias and Wilson 1965. Penzias, A.A.; R. W. Wilson. "A Measurement of the Flux Density at 4080 Mc/s". *Astrophysical Journal Letters* **142**: 1149–1154.
Hafele and Keating 1972. Hafele, J.; Keating, R. Around the world atomic clocks. Predicted relativistic time gains. *Science* **177** (4044): 166–168.
Saburi et al 1976. Saburi, Yamamoto, Harada. High precision time comparison via satellite and observed discrepancy of synchronization. IEEE Transactions vol.25 pp 473-477.
Brecher 1977. Kenneth Brecher. Is the speed of light independent of the velocity of the source? Phys. Rev. Lett. 39 (17), 1051 -1054.
Picasso et la 1977 E. Picasso, J. Bailey, K. Borer, F. Combley, H. Drumm, F.J.M. Farley, J.H. Field, P.M. Hattersley, F. Krienen, F. Lange, and W. von Rüden. (1977). Measurements of relativistic time dilation for positive and negative muons in a circular orbit. Nature Vol. 268, 301.
Newman et al 1978. D. Newman, G.W. Ford, A. Rich, and E. Sweetman (1978). Precision experimental verification of Special Relativity. Physics Review Letters Vol 40 Num 21.
Reasenburg et al 1979. Reasenberg, R. D.; Shapiro, I. I.; MacNeil, P. E.; Goldstein, R. B.; Breiden-thal, J. C.; Brenkle, J. P.; Cain, D. L.; Kaufman, T. M.; Komarek, T. A.; Zygielbaum, A. I.

Viking relativitye experiment. Verification of signal retardation by solar gravity. Astrophysical Journal 234: L219-L221.
Wright 1986. S.E. Wright. Sources and observers in motion I: Time variant analysis and implications to aerodynamic sound. Journal of Sound and Vibration 108(3), 361-378.
Riis et al 1988. E. Riis, S.A. Lee, J.L. Hall (1988). Test of the isotropy of the speed of light using fast-beam laser spectroscopy. Phys. Rev. Lett. 60, pp 81–84.
Krisher et al 1990. T.P. Krisher, L. Maleki, G.F. Lutes, L.E. Primas, R.T. Logan, J.D. Anderson, C.M. Will (1990). Test of the isotropy of the one-way speed of light using hydrogen-maser frequency standards. Phys. Rev. D Vol 42 Num 2. The American Physical Society.
GPS 1992. T. Logsdon. The NAVSTAR Global positioning system. Van Nostrand Reinhold.
COBE 1992. Cosmic Background Explore NASA Goddard Space Flight Center.
Su 2001. Ching-Chuan Su Reinterpretation of the Michelson-Morley experiment based on the GPS Sagnac correction. Europhys. Lett. 56 (2), pp 170-174.
Ashby 2003. N. Ashby Relativity in global positioning systems, Living Reviews in Relativity 2003-2001 Max Planck Institute for Gravitational Physics.
Hubble Space Telescope 2006 Hubble Gallery NASA, ESA, (STScI), and The Hubble Heritage Team (STScI/AURA).
Kramer et al 2006. M. Kramer, I.H. Stairs, R. N. Manchester, M. A. Mc Laughlin, A. G. Lyne, R. D. Ferdman, M. Burgay, D.R. Lorimer, A. Possentti, N. D. Amico, J.M.Sarkissian, G. B. Hobbs, J. E. Reynolds, P.C.C. Freire and F. Camilo. Tests of general relativity from timing the double pulsar. http://arxiv.org/PS_cache/astro-ph/pdf-/0609/0609417v1.pdf
Wright 2008. S E Wright. A new theory forbids material time travel but allows colonization of the universe. Invited talk Café Scientifique, supported by York University, York, UK August 2008.
Wright 2010. S E Wright. Problems with Einstein's relativity, Trafford Publishing.
Gravity Probe B 2011. http://www.sciencedaily.com/ releases /2011-/05/110504150655.htm

The peer reviewed papers below can be accessed on line at piers.org/proceedings, or through the Easy Links indicated.

S.E. Wright [1] "Electromagnetic Sources and Observers in Motion I - Evidence Supporting the EM Propagation Medium for the

Transmission of Light " Paper I of X Progress In Electromagnetics Research Symposium Proceedings. Xi'an, China (March 26, 2010)
S.E. Wright [2] "Electromagnetic Sources and Observers in Motion II - Einstein's Ether-less Relativity Versus Lorentz's Medium Based Theory " Paper II of X Progress In Electromagnetics Research Symposium Proceedings. Xi'an, China (March 26, 2010)
S.E. Wright [3] "Electromagnetic Sources and Observers in Motion III - Derivation and Solution of the Electromagnetic Motional Wave Equation " Paper III of X Progress In Electromagnetics Research Symposium Proceedings. Cambridge, USA (July 5-8, 2010)
S.E. Wright [4] "Electromagnetic Sources and Observers in Motion IV - The Nature of Gravity and Its Effect on the Propagation Medium " Paper IV of X Progress In Electromagnetics Research Symposium Proceedings. Cambridge, USA (July 5-8, 2010)
S.E. Wright [5] "Electromagnetic Sources and Observers in Motion V - A Revised Theory of Relativity " Paper V of X Progress In Electromagnetics Research Symposium Proceedings. Marrakesh, Morocco (March 20-23, 2011)
S.E. Wright [6] "Electromagnetic Sources and Observers in Motion VI - New Motional Optics " Paper VI of X Progress In Electromagnetics Research Symposium Proceedings. Marrakesh, Morocco (March 20-23, 2011)
S.E. Wright [7] "Electromagnetic Sources and Observers in Motion VII - Medium Support for a New Relativity Theory " Paper VII of X Progress In Electromagnetics Research Symposium Proceedings. Moscow, Russia (August 19-23, 2012)
S.E. Wright [8] "Electromagnetic Sources and Observers in Motion VIII - New Relativity Theory Establishes Einstein's Ether-less Aspect of Relativity as Irrational " Paper VIII of X Progress In Electromagnetics Research Symposium Proceedings. Moscow, Russia (August 19-23, 2012)
S.E. Wright [9] "Electromagnetic Sources and Observers in Motion IX - Nature of Gravity" Paper IX of X Progress In Electromagnetics Research Symposium Proceedings. Stockholm, Sweden (August 12-15, 2013)
S.E. Wright [10] "Electromagnetic Sources and Observers in Motion X - Unification of Electromagnetism and Gravity" Paper X of X Progress In Electromagnetics Research Symposium Proceeding. Stockholm, Sweden (August 12-15, 2013)

Acknowledgement

I would like to thank Kelvin Wright for his very considerable help and suggestions, also Nicholas, Jason, Julian and Justine for their help in this work.

Images on Back Page and Front Cover

The images on the back page and front cover of this book are two types of spiral galaxies. M51 and the rotating bar galaxy NGC 1300, respectively. Credit: NASA, ESA, (STScI), and The Hubble Heritage Team (STScI/AURA). The M51 galaxy appears to be representative of spiral galaxies in general. It appears that most well-formed spiral galaxies, without damage through collision or other catastrophic events, tend to have two spiral arms corresponding to two in plane flows of matter being released/jettisoned from opposite sides of a rotating central hub or disk, or from the two ends of a rotating bar as in the galaxy NGC 1300.

The disc or bar material appear to be newer matter (containing younger stars) than the trailing arms, indicating the direction of flow is outwards. Its dynamics appear to be jet like, i.e. material moving out in an orderly radial fashion from the central hub, in two distinct flows. High resolution of the bar galaxy reveals that the central hub material is also rotating. It appears to have two filaments of matter working their way from the centre to the two ends of the bar, where they are released with relatively low velocity, making almost a 90° tip trail, leaving two trailing arms of star producing material.

Rotating in-plane jets, like Catherine wheels, could explain why spiral galaxies are usually flat (disc like) rather than galaxies that are spherical or elliptical. An alternative wisdom is that these flat spiral arms are condensations out of a uniform cloud of energy/mass through some kind of shock wave process. But this mechanism does not explain the flatness of these spiral galaxies and the well formed two jet like structures usually associated with them. Figure 8.4 (d), and equation 8.10 developed for light propagation also seems to apply to matter flows in spiral galaxies, where velocity of light c becomes the jet flow velocity. These photographs are accessible through the Hubble Space Telescope Library (2006).

This is a typical spiral galaxy (M51) representing galaxies not damaged through collision or catastrophic processes. It is thought to be created by some kind of shock wave process through a mass-energy cloud, but this process cannot explain the flatness of spiral galaxies. On close inspection, particularly the bar galaxy (NGC-1300) shown on the front cover, it appears to be generated by two jet flows causing two spiral arms emanating from opposite sides of a central disk, appearing like a rotating Catherine wheel.